Advice and Dissent

ADVICE
AND
DISSENT

SCIENTISTS IN THE
POLITICAL ARENA

Joel Primack

&

Frank von Hippel

Basic Books, Inc., Publishers

NEW YORK

Library of Congress Catalog Card Number: 73-90136
SBN: 465-00090-8
Printed in the United States of America
DESIGNED BY VINCENT TORRE
74 75 76 77 78   10 9 8 7 6 5 4 3 2 1

To our parents

for the values they have taught us.

CONTENTS

PART IV

The People's Science Advisors—Can Outsiders Be Effective?

PART V

Public Interest Science

PART VI

Conclusion

PREFACE

THE common man has never been less in control of his life and livelihood than he is today. Whether confronted by the threat of atomic annihilation or something as trivial as a balky home appliance, almost all of us must place our trust in the hands of the relevant specialists—the people who are ultimately responsible for the design and repair of guided missiles, television sets, and the other complex products of our technological civilization. As our common pool of scientific knowledge increases, the ignorance and powerlessness of each individual increases correspondingly. We can neither smell plutonium nor taste asbestos: the detection of many dangerous materials requires specialized equipment. Each of us must rely upon legions of scientists and engineers for assurance that nuclear weapons and nuclear reactors will not explode accidentally, that adequate fuel will be available to keep us warm and moving and employed, that the chemicals we add to food and water in order to poison our microbial and insect enemies will not poison us as well, and so on through almost every aspect of our lives.

It is evident that the way in which technical experts make their services available to society can significantly affect the distribution of political power. If scientists give government and industry the exclusive benefit of their expertise, they may inadvertently be contributing to the creation of a technological dictatorship in which the uninformed citizen must accept whatever these organizations tell him is in his interest. If, on the other hand, scientists make available to the citizen the information and analyses he needs for the defense of his health and welfare, they can help bring about more open and democratic controls on the uses of technology.

This volume examines how scientists have been carrying out their political responsibilities, in the hope that we may learn both from past mistakes and from past successes. The study grew out of the authors' examination of a number of the most important technological issues that have erupted into national debates during the past few years, including the supersonic transport (SST) project, the antiballistic missile (ABM) program, the safety of the insecticide DDT and the herbicide 2,4,5-T, the decision to take cyclamates out of food, the dangers inherent in the United States' chemical and biological weapons programs, and the safety of commercial nuclear power plants. In studying these issues, we were shocked to find that in every case much of the most important technical advice had been ignored—or worse, publicly misrepresented—by government officials. We were also surprised to find that the final outcome of these controversies was in each case much more influenced by the publicly available information and the public activities of scientists than by the confidential advice given to government officials.

We hope that this book may be useful to two overlapping groups: (1) those actively concerned with the threats posed to society and the environment by an inadequately controlled and explosively developing technology; and (2) scientists who are seeking ways in which they might contribute to the abatement of the multitude of resulting crises.

To the first group, the concerned citizens, we hope to carry the message: *Be skeptical when a government agency cites the opinions of unnamed experts about the safety or necessity of its programs as a substitute for openly discussing the facts and countervailing considerations in a public forum. Demand that the facts and analyses be put on the table and find some independent scientists to check them out with you.* We would also like these people to be aware of how important it is to have independent scientists participating in any public challenge to federal policy for technology. The challengers must understand the issues and be sure of their facts and arguments if they are to be effective and not easily discredited.

To the second group, the concerned scientists, we would like to say: *Writing advisory reports for government agencies is important but not enough. You must be willing to carry your message to the public—by allying yourself with concerned citizens groups, if necessary, and using political and legal pressure to compel government and industry to behave responsibly.* We would also like to make scientists aware that they need not be frightened by the enormous reservoir of expertise available to the federal government. Small numbers of outsiders have had great impact. When the administration in power commits itself to a senseless or dangerous policy, it can no more justify its actions by appeal to its experts than a pilot can justify flying his plane into the ground by quoting the readings on his instruments.

Confidential advice can too easily be ignored. But when a scientist effectively takes his concerns to the public, and these concerns relate to a clear danger to the public health and welfare, then government officials must listen. The challenge to the scientific community—and to the nation at large—is thus to strengthen the government's science advisory system by making it more open and independent, and to encourage and support the "public interest science" movement which has already contributed significantly toward bringing technology under democratic control.

ACKNOWLEDGMENTS

THIS book grew out of a seminar at Stanford University in 1969-1970, led in conjunction with Martin Perl and Robert Jaffe.

The book would never have been written but for the continuing interest, stimulation, and help of our friends and colleagues, especially Joyce Kobayashi, Charles Schwartz, Robert Cahn, Craig Thomson, Jack Uretsky, James Shea, Ralph Lapp, Anne Cahn, Mike Casper, Bob Williams, Gerald Holton, and John Perkins. We are also grateful to our editors, Nina Laserson Dunn and Bill Green, for their faith in the book and for their countless suggestions for its improvement. We appreciate the assistance and cooperation of many other people as well—including some whose names do not appear herein—especially in our investigations of the workings of the federal science advising system.

One of the authors (J.P.) owes a special debt of gratitude to the Society of Fellows at Harvard for both intellectual and financial support.

Portions of our article "Public Interest Science," © 1972 by *Science,* have been reprinted by permission.

PART I

Background

CHAPTER 1

Introduction

A Fairy Tale

Once upon a time there was a vast and beautiful Kingdom. It was endowed bountifully with fertile plains, deep forests, and great waterways. The air was sweet and the waters clear. It was a land of opportunity and many men grew prosperous there.

Many of the greatest Magicians of the Earth were to be found in that Kingdom—for Magicians need opportunities too. They made wonderful inventions, and the Great Men of the Kingdom manufactured these inventions for all the People. The fame and power of the Kingdom grew until it reached every corner of the earth.

But then came a time when there arose Problems. The air over the cities became dark and, when the wind was still, the People coughed. The waters in the rivers and lakes became thick, and the children stopped coming to swim. The food looked more beautiful than ever, but some said that it had become tainted. The quiet places were invaded by noise. And the People lived in fear of death from the skies.

So the People began to talk among themselves: "Perhaps these are evil Magicians. Perhaps their Magic is tainted." But they could not bring themselves to give up the Inventions, so they contented themselves with grumbling: "The King should do something."

The King heard the grumblings and was troubled. He announced: "I will call to the Palace the greatest of the Magicians to advise me." Soon the People saw the Magicians trooping to the Palace and were reassured: "Now the King will find out how to deal with the Problems." And they stopped grumbling.

After the Magicians had spoken to the King in his chambers, the Great Men of the Kingdom came to the palace to learn what he had decided. He told them: "My Magicians tell me that the darkness in the sky, the thickening of the waters, and the other Problems come from your workshops and from the wonderful devices which you make there. They say that you must use the magic more carefully or the Problems will become worse. Therefore I will make some Decrees. . . . "

But when the Great Men heard what the King proposed to do, they threw up their hands and cried out: "But that would mean Ruin! Leave the Problems to us and you will see how soon we will make the Kingdom beautiful and happy again!" The King thought, "These are powerful and successful men. Surely they can deal with the Problems if anyone can." And he announced, "I shall wait."

Perhaps the task was harder than the Great Men had thought or perhaps, being busy, they forgot. At any rate, some years later the Kingdom was even more blighted, and there were more fearful rumors about poisons in the air and the food. The People began to wonder aloud why the King had not solved the Problems. And some of the younger Magicians asked the Great Magicians what their advice to the King had been. They only replied, "We are not free to say." But they continued to troop to the Palace, so most of the young Magicians decided that probably everything would be all right and went back to their studies.

But a few of the young Magicians could not stop worrying. And they began to tell the People that the Problems were getting worse. This, of course, made the Great Men angry. They called these young Magicians "troublemakers," and asked: "How can you possibly think that you know more than the Great Magicians who advise the King?"

How indeed!

The Need for Public Interest Science

This book examines some of the relationships between scientists and the politics of policy-making for technology in the United States. It begins by asking and, we believe, answering the question: Are the advisory efforts of scientists effective in informing the democratic decision-making process? The answer—despite the efforts of the thousands of highly qualified scientists from universities and industry who devote considerable fractions of their time to sitting on technical advisory committees in Washington—is a resounding No!

The reason is simple—and it is the same reason that our country is currently in trouble in so many other areas: virtually all of the advice has gone to the federal executive branch, and with it an almost unchallenged power to make decisions. Administration officials have felt free to ignore or distort technical advice when it hasn't been compatible with their bureaucratic or political convenience. We give examples of how this happens in the case studies in the first half of the book.

Many of these examples deal with events during the Nixon administration, an administration which has become notorious for corruption and the abuse of power. But in many cases the irresponsibility dates back to earlier administrations, supporting our view that these problems stem to a large extent from institutional arrangements and are not peculiar to individuals. In fact, the abuses have a timeless quality—although at some times they may be more blatant and shameless than at others.

In this period of disillusionment, it is important to realize that the assertion of almost imperial powers by the President has led to renewed efforts to defend the checking powers of Congress and the rights of individual citizens. The case studies in the second half of the book illustrate how individual citizens and scientists are working to give those outside industrial and governmental channels access to the policy-making process by making public and comprehensible the information and analyses on which policy must be based.

When dangers become apparent late, the impulse to "look the other way" is strong among the developers and promoters of a technology. If governmental "watch-dogs" are dozing or intimidated, then the public may wait a very long time before corrective action is taken—unless citizen-scientist alliances effectively sound the alarm. It is important that such warning come as early as possible when it is easy to modify technologies or to choose alternatives. This then is "public interest science."

The ordinary citizen or Congressman is often reluctant to become involved in public debates over the uses of technology because he feels that the issues are too technical. The scientist, on the other hand, tends not to speak out because the flavor of the debate is so political. Unless the decision-making process is accessible to the public, however, policies will be decided by those whose careers and livelihoods are affected: bureaucrats and industrialists. People in such positions will ordinarily be the last to acknowledge that something is seriously wrong. It is a self-destructive society which assumes in every case that the interest of such men is identical to the public interest.

The citizen must realize that important political issues are almost always present when a debate of an apparently technical nature bursts into the public arena. Although it may often appear that the partisans involved are asking the public to make determinations in areas where even the experts disagree, the experts are often talking past one another; in reality, the debate revolves around unspoken political questions, such as:

Are you willing to accept sonic booms and increased airport noise in the cause of the "progress" represented by the supersonic transport (SST)?

Inasmuch as nuclear power involves enormous amounts of radioactivity, how much certainty is required that the environment won't be contaminated?

Cyclamates may someday be found to be the cause of tens of thousands of cancer deaths. Are you willing to give up diet drinks to be protected against this uncertain danger?

Do we really want an antiballistic missile (ABM) system which is of uncertain military value—and which may also upset the strategic balance of terror?

In unfamiliar technical areas it is easy to lose sight of such political questions when the "experts" are trying to shoot each other down with technical arguments. And administration spokesmen often play on the self-doubts of citizens who are inclined to "drop out" of such debates. Thus Henry Cabot Lodge (former ambassador to South Vietnam) once told the American public not to involve itself in the debate over the development of the Safeguard ABM system:

This is an argument on which no layman can pass. The judgment of the expert officials whose solemn duty it is to pass on such matters is clearly favorable to ABM.[1]

The concerned citizen must become more sensitive to the political aspects of decision making on technological issues if he is not to be intimidated by such self-serving statements from government officials. It is his future and that of his children which is being decided.

The "expert" must also become more sophisticated. He must become aware of the fact that his actions almost inevitably have political consequences. If he allows government agencies and industry to remain the exclusive beneficiaries of his expertise, he may inadvertently be contributing to the tendency toward a society in which Congress and ordinary citizens are excluded from discussions of policy for technology. Such a fate can only be avoided in our increasingly complex society if scientists are willing to make the information required for participation in these debates more generally available.

To many scientists who have become accustomed to seeing policy making for technology done in private by "expert officials," the idea of involving the public in these issues conjures up disturbing visions. They are concerned about the "Chicken Littles" who will panic at the first suggestion that "the sky is falling." The fact is, however, that when an issue like the SST is taken to the public, the Chicken Littles are not the ones who structure the protracted debate which follows—they will soon be distracted by the next day's sensation. The members of the public who will have an impact on events are the newsmen who will write the stories, the public interest groups who will decide whether or not to commit their limited resources to the debate, the lawyers and judges who will identify and decide on the legal issues, the state and local officials who may feel that they must take initiatives to protect the health and welfare of their constituents, and the Congressmen and their staffs who must decide whether investigative or legislative action is called for. If important issues of public policy cannot be discussed productively by these groups, then there is little hope for democracy. In fact, the evidence from our case studies would appear to indicate that misleading statements issued by government spokesmen endanger the integrity of the debate far more often than does irrational behavior on the part of the public.

Some scientists worry lest their involvement in public interest activities invite retaliatory funding cuts and restrictions—this at a time when scientists are already concerned over the erosion of their security and their independence. The possibility of retaliation must be taken seriously and measures taken to combat it. One reason why science is currently in trouble, however, is that the public feels that the technology which science has made possible has been exploited in an irresponsible way. As a result, scientists have come to be seen as amoral technicians much like the rocket engineer caricatured by Tom Lehrer:

Once the rockets are up,
Who cares where they come down?
That's not my department,
Says Wernher von Braun.[2]

If public interest science activities can help bring sanity to the direction of technology, the faith of the public in scientists may to some degree be restored.

These considerations are timely because the renewed public awareness of what happens when the exploitation of technology is left effectively under the control of special industrial and governmental interests has led to an increased readiness within the scientific community to undertake serious commitments to public interest activities. But can such efforts be successful? The case studies in the second half of the book constitute remarkable evidence that they can.

Plan of the Book

Surprisingly few scientists and even fewer concerned citizens have ever followed in any detail the activities of the "experts" during a national debate over a major technological issue such as the SST development project. Nonparticipants hear executive branch officials, political figures, and newspapers cite various experts—but seldom do we hear in useful detail what the experts actually said and to what effect. In the next chapter, therefore, we lay out the story of the scientists in the SST debate. This chapter describes both the roles of the "insiders"—the technical experts who advised Presidents Kennedy and Nixon, the Federal Aviation Agency, and the Department of Transportation—and the "outsider" scientists who became concerned about the environmental impact of the SST and took their concerns directly to the public and to Congress. The development of the controversy followed a pattern which has become rather typical. First was the long period during which an executive branch agency nursed the monster, blithely ignoring advisory reports about the adverse consequences that could be expected to accompany its maturity; then the first expression of public concern by an independent scientist; the development of the issue into a political controversy in which Congress started to feel political heat and the leadership took sides; then the debate in which the executive bureaucracy tried to mislead Congress and the public by invoking its expert advisors; and finally Congress's decision.

Having set the scene with an example of a controversy over technology, we develop in the remainder of the book two main themes: the limitations on the effectiveness of the government advisor, and the importance of the public interest scientist in keeping the policy-making process "honest." Despite the almost catastrophic decline in respect for the federal executive branch in the past few years and the attendant rise in citizen activism, it seems to us that the basic attitudes which got the nation into this trouble are still very much in evidence and await only a return of government to some form of "normalcy" to reassert themselves. This book attempts to combat the views that "the only effective way to influence federal policy is by working on the inside" and, conversely, that "you can't fight City Hall." We present the case for the opposite views—at least in the area of federal policy for technology.

In Parts I-III we show how, despite the federal executive branch's legions of science advisors, when it comes to actual decision making the Emperor very often runs about in the buff. But, as in the story of the Emperor's new clothes, the public is deceived by the mere existence of the process into believing that, if something appears missing in the Emperor's garb, something must be wrong with themselves. The public relations artists in the story have true descendants in the modern government spokesmen who cite the great distinction of the government's advisors to intimidate the concerned citizen into disbelieving his own—often very accurate—perceptions.

In order to help immunize our readers to this tactic, we will inoculate them with several case studies illustrating the devices by which the reputation of the executive branch's science advisors has often been used to buttress the very policies which they had opposed.

Lest these disheartening stories about the abuse of the executive's science advisory system convince the reader that all advisors should disassociate themselves from the executive branch and that those who don't must obviously be "prostitutes," we include (Chapter 3) a discussion of what the legitimate purposes of the advisory system are. We also discuss (Chapter 9) the various attempts which have been made in recent years to make the executive branch advisory system more open and less vulnerable to subversion. The focus throughout is on the "confidentiality" of sensitive advisory committees which has made it easy to suppress and misrepresent their reports. The obvious remedy is to open up the advisory committee membership and reports to outside criticism—and legislation to this effect has recently been passed. It seems unlikely that this legislation will be successful in achieving its objective, however, unless it is understood and supported by both concerned citizens groups and the scientific community.

Even if these reforms are successful, there will still be a need for a pluralistic advisory system to give groups outside the executive branch the independent information and analyses necessary to judge the issues. The current practice of public interest science represents the beginnings of what will hopefully grow into a structure comparable in resources to the executive branch's science advisory establishment. Already the present ad hoc and part-time public interest science activities have had great impact on events. In Part IV we present some examples of this impact in areas ranging from pesticide regulation to nuclear reactor safety. If it were not for the small number of public interest scientists who have taken these issues to the public, our society might have stumbled deeper into several technological Vietnams.

With these examples in mind, we discuss (Part V) the strengths, limitations, and possible future of public interest science. At the moment scientists are undertaking this activity in many different settings: in universities, scientific societies, and public interest groups. These public interest activities complement efforts to increase the capabilities of Congress to deal with technology. In 1973 Congress set up an Office of Technology Assessment. At the same time, several professional societies established year-long fellowships for scientists to work on Capitol Hill. We speculate in Part VI on the possibility that such initiatives may represent the beginnings of a new and healthier relationship between scientists and society. For the reader who wishes to be reminded of the identity of the major science advisory organizations cited, brief descriptions are provided in the Appendix.

NOTES

1. Henry Cabot Lodge, "A Citizen Looks at the ABM," *Readers' Digest,* June 1970, p. 63.

2. Tom Lehrer "Wernher von Braun," on *That Was the Year That Was* (Hollywood, Calif.: Reprise Records, 1965). Used by the permission of Tom Lehrer.

CHAPTER 2

The Supersonic Transport: A Case History in the Politics of Technology

Never in my experience has the "big lie" technique, popularized by Adolf Hitler's propaganda minister in World War II, been used more effectively to describe a needed program of research and development. . . .

It was not only amazing but down-right frightening to see the number of prominent scientists who were willing to lend their names to far-fetched and hypothetical possibilities. . . .

The scare techniques used against the SST are similar to the ones that were used by some of the same people to oppose the A-bomb tests in Bikini in 1946, the development of an H-bomb in 1949, and even to such beneficial humanitarian projects as building a dam across the Colorado River in the Grand Canyon.[1]

—Senator Barry Goldwater

In 1970 and 1971 a major national debate raged in the United States over the federally funded project to develop a commercial supersonic transport (SST)—a new aircraft which could carry passengers long distances at speeds greater than that of sound. Senator Goldwater's remarks testify to the intensity of that debate and to the great impact of scientists in it—although his assessment of the nature of their impact is surely idiosyncratic.

From 1963, when President Kennedy committed the federal government to

10

the SST project, until 1971, when Congress finally killed it, nearly a billion dollars were spent on SST development and design. During the course of three Presidential administrations, the project successfully survived a number of technological setbacks and adverse governmental reviews. What finally halted the SST was the growth of widespread public opposition based on a popular impression that the taxpayers' money was being wasted building an economic white elephant whose operation would constitute a serious public nuisance.

In this chapter we will trace the contributions of scientists as advisors to the government in the repeated reviews of SST development and as advisors to the nation as a whole in focusing attention on the aircraft's economic problems and potential for environmental degradation.

Sonic Boom, Engine Noise, and Economics

Two of the major environmental problems associated with the SST—sonic boom and engine noise—were already generally recognized within the government by the time President Kennedy made his decision to go ahead with the project. These problems were considered in the feasibility studies which were conducted or funded by the government during 1960-1963, they were discussed publicly in Congressional hearings on the subject during the same period,[2] and they were taken explicitly into account in the design objectives specified for the SST in the proposal which Kennedy sent to Congress in 1963.[3]

Any object traveling through air faster than sound produces a supersonic shock wave, much like the bow wave of a motor boat. When this shock wave reaches the ground it is felt as a loud, explosive noise: the sonic boom. The SST's sonic boom was to be limited, according to President Kennedy's proposal, to an overpressure during acceleration of less than 2 pounds per square foot (psf) and during cruise of less than 1.5 psf. The hope was expressed that the public might tolerate booms of these intensities. A sonic boom of one psf was, according to the proposal, expected to be "acceptable" to the public. "Some scattered public reaction" was expected at 1.5 psf, and "probable public reaction—particularly at night," was expected at 2 psf. Sonic booms with intensities of 2.5 psf were likened to "close range thunder or explosion" to which the proposal, not surprisingly, expected "significant public reaction."[4] The acceptability of more intense booms was not even considered. (It is important to realize that the sonic boom from a supersonic aircraft is felt on the ground in a "boom carpet" tens of miles wide extending over the *entire* supersonic flight path of the plane, not just when it accelerates past the speed of sound or when it is flying below its cruise altitude.)

President Kennedy's proposal also included the design objective that the engine noise of the SST be no greater than that of "current international subsonic jet transports."[5] The noise of subsonic jet operations was already disturbing populations even at considerable distances from metropolitan airports, so it was recognized that particular care would be required in the design of the powerful engines required for the SST.

The other design objectives, as Kennedy presented them to Congress, called for an aircraft weighing 350,000 pounds, with a payload of 35,000 pounds and a range of 4,000 statute miles. Its cruise speed was to be better than 2.2 times the speed of sound,[6] considerably faster than that of the Concorde, the SST being developed jointly by Britain and France. As a result, the aircraft surface would be heated up by friction to such high temperatures that aluminum, the standard material used in subsonic aircraft and in the Concorde, would have to be replaced by titanium, a metal both more expensive and more difficult to work. Finally, the SST was to be able to operate from existing international airports at operating costs comparable to subsonic jets.

These last requirements were essential if the SST was to compete successfully with existing airplanes. There was little doubt that the technical objectives could be met, but whether they could be met in an economically competitive aircraft was the crucial question. Even the manufacturers who were vying for the federal contract were unable to present more than a marginal case that the SST would compete successfully with large subsonic jets. In fact, the Stanford Research Institute, whose market estimates were used in President Kennedy's proposal, had come to the flat conclusion that "there is no economic justification for an SST program."[7]

The Political Decision

The initial advocates of a federally funded SST project were the aircraft industry, the federal agencies concerned with aviation, and the U.S. Air Force. The program obtained full federal commitment as a result of a general conviction in these circles that the supersonic transport represented the next inevitable advance in commercial aviation and the fear that Soviet or Anglo-French domination of the SST market would be a terrific blow to American prestige, the U.S. balance of payments, and the competitive ability of one of the country's strongest industries.

In early June 1963 a special review committee composed of administration officials and headed by Vice-President Lyndon Johnson submitted recommendations to President Kennedy for an American supersonic airliner project. The British and the French had already three years before agreed to collaborate on developing their own SST, the Concorde, and their effort was being taken very seriously, particularly after Pan American World Airways announced that it had acquired options on six Concordes. Within a few days after receiving the review committee's report, President Kennedy announced, in a commencement speech at the Air Force Academy, his decision to proceed with the project, adding:

"The Congress and the country should be prepared to invest the funds and effort necessary to maintain this Nation's lead in long-range aircraft."[8] The circumstances of the announcement allowed it to serve another purpose as well: it reassured the Air Force that the technology of sustained supersonic flight by large aircraft would be developed despite the administration's recent cancellation of the B-70 supersonic bomber. (Sitting perpetually in an underground limbo next to a missile silo waiting for doomsday seemed to the Air Force a far cry from the "wild blue yonder.")

From this brief description of the origins of the United States' SST project, it is evident that the dynamics are analogous to those which have become classic in the strategic weapons race. Other nations had responded to the American dominance of the long-distance subsonic transport market by planning to develop a faster aircraft. The Americans then felt compelled to rise to this challenge by developing an even faster aircraft. The government officials involved appeared to realize that these developments were technologically premature and might well result in less economical air transportation and a substantial degradation of the human environment. But they felt that there was no way to escape the logic of international competition.

In view of the many risks and uncertainties involved in the enterprise, however, President Kennedy tried to delineate in his proposal certain decision points in the development program at which the project could be redirected or even terminated. In his message to Congress later in June 1963, he described the major dangers as follows:

1.   That technological problems cannot be satisfactorily overcome,
2.   that a supersonic transport will not have satisfactory economics, [or]
3.   that sonic boom overpressures will result in undue public disturbance.[9]

In retrospect, this list appears to have been prophetic.

"We Are All-Out For Economics Now"

It had been anticipated in Kennedy's proposal that the design competition phase (in which manufacturers bid for the government contract) and the detailed design phase of the SST project would be completed by 1965. In fact, the Federal Aviation Administration (FAA), which had been made responsible for overseeing the project, did not accept a final design from the Boeing Company until 1969, during the Nixon administration. Boeing's variable-geometry ("swing-wing") design had finally been chosen in 1967 over Lockheed's fixed-wing design. The FAA hoped that with this design a moderate-size SST with a tolerable sonic boom might be economically viable. But after another year of trying to perfect the design, Boeing finally admitted to the FAA that the swing-wing idea was impractical: the machinery necessary to hold and move the wing was simply too heavy. A choice was therefore necessary among a less economical plane of the original size, a larger plane with a more intense sonic boom—or the cancellation of the project.[10]

The FAA opted for a large plane. The aircraft grew to have a gross design weight of 750,000 pounds—as great as that of the Boeing 747 jumbo-jet and more than twice the maximum weight which had been set as a design objective in Kennedy's original proposal. The expected average sonic boom overpressure grew correspondingly to 2 psf during cruise and 3.5 psf during acceleration—even greater than the sonic boom intensity that Kennedy's original proposal had compared to "close range thunder or explosion." An (anonymous) administration official put the new FAA position succinctly in an interview with the *New York Times*: "We are all-out for economics now and to hell with the boom."[11]

The decision on the SST engine went much the same way. A 1960 report on the SST by the National Aeronautics and Space Administration (NASA) had concluded:

It is obvious that noise considerations will have an important bearing on the choice of structure, the power plant, the aerodynamic configuration, and the operating practices. These noise problems should thus be considered early in the design stage of the airplane.[12]

A few years later, however, when SST designers were fighting to pare every extra pound off the aircraft design, this admonition had been forgotten. Although the engine design competition could readily have been arranged to permit a direct comparison of noise levels, environmental considerations were pushed so far into the background that noise was forgotten as a serious consideration in the selection of the SST engines.[13] The result was that Pratt and Whitney's relatively quiet duct-burning turbofan design was rejected in favor of General Electric's afterburning turbojet design. The General Electric engine would have given the SST a sideline noise far greater than that of any modern jet aircraft.

The Citizens League Against the Sonic Boom

Even though the government had given the sonic boom problem a low priority, it proved to be difficult to ignore. In 1964 the FAA conducted a major test of public acceptance of sonic booms. In this test, the 300,000 citizens of Oklahoma City were subjected to booms averaging 1.3 psf overpressure eight times daily for five months. At the end of the test only 73 percent of the Oklahoma City residents polled felt that they could learn to tolerate booms of this intensity, even during working hours. More than 15,000 persons filed complaints, and almost 5,000 filed damage claims for broken glass and plaster which resulted in compensatory payments and awards totaling $218,000.[14]

As a result of the Oklahoma City test and other data, government science advisors were becoming increasingly skeptical of the possibility of commercial supersonic flight over populated areas.[15] The FAA remained persistently optimistic, however. The director of the SST project, Gen. Jewell C. Maxwell, stated in 1968: "We believe that people will come to accept the sonic boom as they have the rather unpleasant side-effects of other advances in transportation."[16] And the FAA continued to base many of its economic analyses and market assessments upon the assumption that the SST would be permitted to fly supersonically over land.

In 1967 the first serious attempts were made to take the SST sonic boom issue to the public—mainly as a result of the efforts of one remarkable individual, Dr. William A. Shurcliff, a soft-spoken, white-haired Bostonian of refined and gentle appearance. During the Second World War Shurcliff had served as an administrative assistant to Vannevar Bush, Director of the Office of Scientific Research and Development, acting as the office's liaison to the Manhattan Project (which developed the atomic bomb). Later he worked at the Polaroid Corporation, and for the next ten years he assisted in the administration of the Harvard-MIT Cambridge Electron Accelerator. He has been retired since 1973.

Early in 1967, Shurcliff decided to try to organize and strengthen public opposition to the sonic boom, his interest in the issue having been aroused by an article in the *Bulletin of the Atomic Scientists* by the Swedish aeronautical expert and SST opponent Bo Lundberg.[17] Somewhat later, the *New York Times* published a letter expressing opposition to the SST from John T. Edsall, an eminent Harvard biologist. Shurcliff went to see Edsall and, after determining that no organized opposition to the SST existed, they founded the Citizens League Against the Sonic Boom (CLASB) in March 1967.

During the period from 1967 through 1971, Shurcliff devoted almost all of his spare time to the job of running the League. He recalls that he spent four or five hours most weekday evenings at it—and most weekends as well. At first he hired a secretary, but soon he found that it was faster for him to compose his letters and press releases at the typewriter in his home office and send them off as they came out. He had a similar experience with a rented addressing machine that he used for addressing his frequent newsletters to the membership of CLASB, which soon grew to number some 4,000. After having continual problems keeping the rented machine adjusted, Shurcliff built his own addressing machine in his attic. It is simplicity itself, its parts including assorted pieces of wood, a couple of hinges, some rubber bands, and an old rubber bicycle handlebar grip. With the assistance of his son, he can use this ingenious device to address 4,000 newsletters in four hours.

In the course of his campaign against the SST, Shurcliff distributed more than a score of press releases to some 200 newspapers. These releases received good coverage (appearing in an average of five newspapers a day in 1967 and 1968). This was no doubt partly because CLASB was the only group distributing such material at the time, but it was also because the press releases were generally accurate and well written. Shurcliff's "bang zone" maps, showing typical areas of the United States and the Atlantic Ocean which would be subjected to regular sonic booms, were widely reproduced. Shurcliff also compiled the *SST/Sonic Boom Handbook*, which was later expanded and published as a Ballantine paperback in 1970 in collaboration with the newly formed Friends of the Earth organization. The *Handbook* is a model of informative and responsible advocacy. More than 100,000 copies were sold to the public. Shurcliff bought an additional 10,000 copies for CLASB at sixteen cents each and mailed them to Congressmen, airline officials, and whomever else he thought should be confronted with the powerful arguments against the SST.

One of the principal factors in the success of Shurcliff's fight against the SST was his considerable faith in people—as well as in the rightness of his cause. For example, Shurcliff and a few other SST opponents put together $7,000 for two half-page advertisements in the *New York Times* attacking the SST and inviting readers to join CLASB. Among the responses was an anonymous donation of $10,000.

Shurcliff feels that the personal touch is very important. Whenever he mailed out some special material to a member of CLASB, he made sure to write a personal note at the top of the first page in red pencil "to make sure that they don't miss it." When asked if he stopped sending mailings to CLASB members who did not contribute, Shurcliff replies emphatically: "Never!" He then explains that some "poor" members who have contributed nothing at all helped the effort in other ways by frequently writing their Congressmen or by forwarding him useful clippings from their local papers. Shurcliff's dedication and personal loyalty to the CLASB membership was well reciprocated. Once or twice a year as required he would add a note at the bottom of a newsletter: CLASB NOW NEEDS MONEY. Invariably the response would be on the order of $10,000. The funds raised through CLASB provided the majority of the financial support for the effort against the American SST, and CLASB provided about one-third of the support for the anti-Concorde effort in Britain.

More than anyone else, Shurcliff deserves the credit for having made it impossible to fly SSTs over the United States.

President Nixon Reviews the SST Program

January 15, 1969, was the final deadline for Boeing to submit a revised SST design to the FAA. The failure of the swing-wing idea and Boeing's continuing design difficulties had resulted in a delay of several years, during which substantial opposition to the sonic boom had developed. When President Nixon took office in January 1969, he therefore announced that he would reassess the SST program. He immediately commissioned two comprehensive reviews of the SST's economics and environmental impact. One of these was undertaken by the sub-cabinet-level interdepartmental ad hoc SST Review Committee and one by a panel of outside technical experts headed by a member of the President's Science Advisory Committee, Richard Garwin.

The SST Review Committee members included Undersecretary of the Interior Russell Train; Hendrik Houthakker of the Council of Economic Advisors; Lee DuBridge, Director of the Office of Science and Technology and science advisor to the President; and other officials of similar stature. The committee thus appears to have been as high-level and broadly based a working committee as one could expect to have assembled within the executive branch. It divided into working panels which considered different aspects of the issue. A month later the panels returned with reports which were highly unfavorable to the SST project.[18]

The Panel on Balance of Payments and International Relations concluded that the threat of foreign competition, which had originally triggered the American SST program, was not materializing:

The viability of the Concorde is very much in doubt—particularly because of landing and take-off noise, range limitations and prospective high operating cost per seat mile.[19]

Based on the other panel reports, the United States' SST seemed to be in similar trouble. Thus, the Economics Panel reported that there was "a large element of doubt" on the subject of the SST's ability to compete economically with subsonic jets.[20]

Perhaps the most important conclusions, however, were those of the panel studying the impact of the SST on the human environment, which reported that "all available information indicates that the effects of the sonic boom are such as to be considered intolerable by a very high percentage of people affected."[21] The same panel also concluded that

Noise levels associated with SST operations will [be such] that significant numbers of people will file complaints and resort to legal action, and that a very high percentage of the exposed population will find the noise intolerable and the apparent cause of a wide variety of adverse effects.[22]

The environment panel's report also mentioned the possibility that the water vapor in the SST exhaust gases might have serious effects on the upper atmosphere and weather.

Finally, the panel studying the impact of the SST program on the aerospace industry concluded that the impact was "difficult to assess, but it appears small,"[23] supersonic technology having already been developed for military applications.

The other comprehensive review commissioned by President Nixon resulted in a report (the "Garwin Report") which was even more unfavorable. Unlike the government officials on the SST Review Committee, this panel of technical experts capped their criticisms of the American SST program with a very explicit recommendation:

We recommend the termination of the development contracts and the withdrawal of Government support from the SST prototype program.[24]

On September 23, 1969, half a year after receiving these reports, President Nixon announced his decision to go ahead with the program. He gave as his primary reason the one President Kennedy had given six years before: "I want the United States to continue to lead the world in air transport."[25] As usual, the executive branch attempted to keep the unfavorable reports on the SST confidential. Nixon's only concession to SST critics was a statement, issued through Transportation Secretary John Volpe, that the SST would not fly at supersonic speeds over the United States.[26] But there were some indications that the administration continued to believe that the SST would eventually be allowed to fly domestic routes.[27] The government continued to equivocate on the matter until 1972 when the FAA issued a rule prohibiting commercial jets from producing sonic booms over land.

All three of the eventualities which President Kennedy had listed as grounds for termination or redirection of the SST program had come to pass: the technological problems involved in making a small, quiet, economical aircraft had not been overcome; the proposed SST did not have satisfactory economics; and its sonic booms would be intolerable to the public. Yet President Nixon gave the program his blessing. Whatever the reasons for his decision to continue to support the SST project, President Nixon's announcement effectively terminated debate on the SST within the executive branch. The focus of the national debate shifted to Congress.

The Battle for Congress

The appropriations for the SST program which President Nixon had requested in 1969 were passed by Congress within a few months by lopsided votes which differed little from those of 1966 or 1967 (no additional appropriations had been requested in 1968).[28] The vote was essentially unaffected by the fact that the SST Review Committee documents had by then become available as a result of the strenuous efforts of Representative Henry Reuss (D.-Wisc.).[29] (The Garwin Report remained secret until long after the end of the SST debate.)

How can we understand the lack of impact on Congress of these documents and other adverse information? The answer seems to be that the SST project had become part of the intricate network of political arrangements by which Congressmen protect the interests of their corporate constituents. The stability of the voting pattern was enhanced by the fact that the chief Senate proponents of the SST were Senators Warren Magnuson and Henry Jackson, both Democrats of Washington—not coincidentally the home of the Boeing Company. These Senators chaired committees and subcommittees which allocate and pass upon billions of dollars of program funding and were therefore in a much better position to do favors, collect debts, or retaliate against other Senators than the much less advantageously positioned Senators who led the opposition: Senators William Proxmire (D.-Wisc.), J. William Fulbright (D.-Ark.), and Gaylord Nelson

(D.-Wisc.). After the 1967 Senate vote on the SST appropriations, Magnuson bragged to a newsman: "What'd Proxmire get? Nineteen votes? I could have had half of those if I'd needed them."[30]

It was only half a year after the "business as usual" 1969 vote, however, that the House of Representatives barely passed the SST appropriations by a vote of 176 to 172. Then, on December 3, 1970, the Senate voted down the appropriations by a vote of 52 to 41. The SST died officially four months later, when both Houses agreed to terminate the project. In the intervening year, a full-scale national debate had developed, making the SST program one of the major political issues of the Congressional election year 1970. An aide to Senator Magnuson later tried to explain the Senate reversal:

[Magnuson and Jackson] called upon every Senator they thought they could influence this time. They called. They cajoled. They persuaded. They arm twisted. They did everything they could. But you can't push something down the throats of the Senate. The SST became a big national issue, and it was just beyond the power of the Senators to turn around.

Vote trading and arm twisting is effective when the issue is not that big, when it isn't a glaring national issue. But it doesn't work when you've got the full focus of national attention on it. Then the pressure is on, as Senators will say, to "vote right."[31]

By 1969 Shurcliff, Representatives Reuss and Yates, Senator Proxmire, and others had made the opposition to the SST visible. The newspapers were eager to feature any new developments in the debate, and national political figures became involved. Senators Edmund Muskie (D.-Maine) and Charles Percy (R.-Ill.) joined the political mavericks in the Senate opposition, and New York's Governor Nelson Rockefeller vowed to keep the SST out of his state. In Washington, the lobbyists of major conservation and environmental groups worked with sympathetic Congressional aides to provide arguments with which the aides could sway their bosses and which the Congressmen could later use to explain their change of mind. And in many states environmentalists injected the SST into the 1970 Congressional campaigns. Many Senators who had previously voted for SST appropriations were boxed into a corner by their opponents and forced to make anti-SST statements to satisfy their constituents. It has been suggested that many of these Senators may have comforted themselves with the thought that their votes would not be needed by the pro-SST forces, since Proxmire's previous attempts to stop the SST had always been defeated overwhelmingly.[32]

Selling the SST to Congress

Although many political currents and countercurrents flowed in the national debate over the SST, the Congressional hearings which were held to establish the facts of the SST's economic prospects and environmental impact were crucial. It was in these hearings that most of the information which was carried by the news media was developed.

In these hearings, administration officials tried to present the strongest possible case for the SST; and in the process they generally overstated its advantages and understated its disadvantages. For example, the Department of Transportation, defining aircraft "productivity" as cruising speed times the number of seats, claimed that the SST would be twice as "productive" as a Boeing 747.[33] This comparison obviously ignored the SST's comparatively short range, negligible cargo capacity, and high fuel consumption per seat-mile compared to subsonic commercial jets, as well as the larger proportion of time each trip that the SST had to spend on the ground. (Although the SST could *fly* three times as fast as a conventional jet, it would take just as long to taxi, load and unload passengers, and be serviced.) The Department of Transportation also stressed the balance-of-payments advantages of exporting SSTs instead of importing Concordes, but it refused to consider other, probably equally serious balance-of-payments consequences of developing the SST.[34] Meanwhile, advertisements placed by the lavishly funded pro-SST lobby prematurely proclaimed the imminent entrance of the Soviet supersonic airliner into commercial service.[35]

In speeches and in Congressional testimony, the new FAA Administrator, John Shaffer, insisted that the SST's sonic boom is "not destructive," despite readily available evidence to the contrary—for example, the damage caused in the Oklahoma City sonic boom acceptability tests. Summing up the administration's view of the SST's environmental impact, William M. Magruder, Director of the Department of Transportation's new Office of Supersonic Transport Development, stated:

According to existing data and available evidence, there is no evidence of likelihood that SST operations would cause significant adverse effects on our atmosphere or our environment. This is the considered opinion of the scientific authorities who have counseled the government on these matters over the past five years.[36]

It is very difficult to see how this statement can be squared with the technical advice available to the Nixon administration—for example, that summarized in the SST Advisory Committee report.

Those officials and technical experts who had opposed the project within the administration were generally silent. In April 1970, however, Senator Proxmire wrote to the members of President Nixon's SST Advisory Committee to ask them whether they had learned anything in the intervening year to change the views which they had expressed in their report. With one exception, they replied that their views had not changed substantially.[37]

The exception was Lee DuBridge, the President's science advisor. In March 1969 he had written to the chairman of the SST Review Committee:

Granted that this is an exciting technological development, it still seems best to me to avoid the serious environmental and nuisance problems and the

Government should not be subsidizing a device which has neither commercial attractiveness nor public acceptance.[38]

In April 1970 DuBridge replied to Proxmire's question:

Needless to say, the President has a broader view of the whole problem after he has studied all the facts and opinions which have been brought to his attention. Thus, while each of the several of us may have, from our own restricted points of view, recommended against further federal involvement in the SST project, I, for one, believe that the President, in taking a more comprehensive view than any of us could have, came to a sound decision. . . . The President recognizes, as we pointed out, that there are still technological and environmental problems to be solved. But he has the faith, which I now share, that the ingenuity of the American industrial system can eventually solve these problems satisfactorily.[39]

After assuming this undignified posture, the former president of Cal tech hardly had to explain to Representative Sidney Yates (D.-Ill.): "Congressman, I am a soldier. The President has made up his mind, and I am going to support the President's decision."[40]

In September 1970 DuBridge was replaced as Presidential science advisor by a physicist from Bell Laboratories, Edward David, Jr. David continued DuBridge's refusal to release the Garwin Report. He also actively campaigned for the SST: in December 1970 he issued a pro-SST statement, co-signed by thirty-four prominent scientists and engineers, which contended that the Senate vote against the SST "represents the wrong approach in dealing with new technology."[41]

Testimony against the SST

A serious break in administration ranks occurred when Russell Train, now Chairman of the Council on Environmental Quality, appeared before Senator Proxmire's Joint Economic Committee in May 1970, accompanied by Dr. Gordon J. F. MacDonald, a geophysicist and fellow Council member. Train emphasized in his testimony the seriousness of the SST airport noise problem and also discussed the possible impact of pollutants from the high-flying aircraft on the Earth's stratosphere. He characterized stratospheric pollution as

. . . a potential problem which has not received the attention it deserves. The supersonic transport will fly at an altitude between 60,000 and 70,000 feet. It will place into this part of the atmosphere large quantities of water, carbon dioxide, nitrogen oxides and particulate matter.
    A fleet of 500 American SSTs and Concordes flying in this region of the atmosphere could, over a period of years, increase the water content by as much as 50 to 100 percent. . . . First, [this] would affect the balance of heat in the entire atmosphere leading to a warmer surface temperature. . . . Second, water vapor would react so as to destroy some fraction of the ozone that is resident in this part of the atmosphere. The practical consequences of such a destruction

could be that the shielding capacity of the atmosphere to penetrating and potentially highly dangerous ultraviolet radiation is decreased. . . . Finally, the increased water content coupled with the natural increase could lead in a few years to a sun shielding cloud cover with serious consequences on climate.

Clearly the effects of supersonics on the atmosphere are of importance to the whole world. . . . The effects should be thoroughly understood before any country proceeds with a massive introduction of supersonic transports.[42]

According to a Proxmire aide, Train's testimony gave the "stamp of seriousness" to concerns about the potential impact of a fleet of SSTs on the stratosphere—concerns that had previously been dismissed as far-out scare stories. The exact nature and extent of this problem remains uncertain—recent work suggests that the nitrogen oxides problem may be much more serious than it was thought to be in 1969 and the water vapor problem perhaps less important—but it seems clear that Train's final conclusion quoted above has lost none of its force.[43]

Train was the only important administration official who publicly gave testimony damaging to the SST project. However, another witness who appeared in opposition to the SST had been a confidential advisor to the Executive Office on the matter: Richard Garwin.

Garwin did not volunteer to testify on the SST, but was invited to appear before several Congressional committees after DuBridge, the President's science advisor, had publicly mentioned the Garwin Report. In his testimony, Garwin detailed how each time that Boeing had failed to meet the specifications of the SST contract, the FAA obligingly issued new ones specifying a considerably less desirable airplane. He summarized:

The development contract won by Boeing on the basis of the swing-wing design and requiring the prototype to be very close to the actual version, as well as to have outstanding takeoff and landing characteristics, has been successively modified to the point at which it is problematical whether the SST will fit on existing airfields, and to a point where the airport noise is far beyond the maximum acceptable for jet aircraft now.[44]

Perhaps Garwin's most widely quoted observation was that "at 125 PNdB of airport noise, the SST will produce as much noise as the simultaneous takeoff of 50 jumbo jets satisfying the 108 PNdB subsonic noise requirement."[45] Once he had thus made public his opposition to the program, Garwin wrote letters to the editors of newspapers and appeared on television; he also went in person to present the arguments to individual Congressional supporters of the SST—and was reportedly quite effective.[46] Garwin's testimony and the administration's increasing emphasis on the jobs which the SST program would provide deepened the suspicion held by many observers that the SST project had become, to a great extent, an expensive form of welfare for the depressed aerospace industry.

Laurence Moss, a young engineer on the staff of the National Academy of Engineering, was another technical expert who played an important role in the SST debate. Moss had become disenchanted with the SST while serving as a White House fellow assigned to the Department of Transportation. He then participated as an individual in the effort to stop the SST project both as an organizer and as an expert witness before Congressional committees.

In early 1970, Moss advised Senator Muskie on the SST issue. When Muskie's staff was approached by a wealthy schoolteacher who wanted to make a major contribution to the anti-SST campaign, they therefore put him in contact with Moss. With this financial backing, Moss was able to bring the anti-SST groups together to organize the Coalition Against the SST.[47] The Coalition was very effective in organizing lobbying and in the popularization and wide distribution of statements and information which had been prepared by SST critics such as Shurcliff. One of its coups, in collaboration with two of Senator Fulbright's aides, was to persuade fifteen prominent economists, ranging in philosophy from Milton Friedman to John Kenneth Galbraith, jointly to publish statements explaining their opposition on economic grounds to the SST.[48] After its founding contribution, the Coalition received a substantial fraction of its funding from the Citizens League Against the Sonic Boom.

Moss excelled in translating numbers into tangible quantities. In Congressional testimony he presented the noise problem as follows:

The disturbance at 1 mile from a subsonic jet is about the equivalent of the disturbance at 15 miles from the SST. . . . In other words, the "sideline noise" implied by SST proponents to be an airport, not a community problem, will be highly objectionable at distances of over 15 miles from an airport intensively used by the SST.[49]

This point was reemphasized in October 1970, when the Federation of American Scientists, a Capitol Hill lobbying group dominated by prominent scientists, distributed to every Senator a set of maps which showed the Senators that all or most of the metropolitan areas of New York City, San Francisco, Seattle, Honolulu, Anchorage, Boston, and Los Angeles would be affected by the SST engine noise.

Moss also drew attention to the extravagant use of fuel by the SST:

An SST with 300 seats . . . consumes 0.33 pounds of fuel per seat-mile. This is about twice the fuel consumption per seat-mile of the Boeing 747. . . . A fleet of 500 SSTs, each flying the equivalent of three transatlantic round trips per day, will burn about 1.2 billion pounds of fuel per day. . . . This amount of fuel, by the way, is almost equal to the fuel consumed each day by all 105 million motor vehicles in the United States.[50]

George Eads, then a young assistant professor of economics at Princeton specializing in aircraft and airline economics, was another expert who participated effectively in the campaign against the SST. When the Transportation Department sent a map to each Congressman's office showing the amounts that his state could expect to receive in SST subcontracts, it was Eads who prepared a map for distribution by the Coalition Against the SST which showed that all but a few states would contribute more in taxes than they would receive in subcontracts. Eads also pointed out in Congressional testimony that Congress had once before been asked to fund a development project for a commercial air transport.[51] This was just after World War II, when the British government-financed development of a commercial subsonic jet appeared to pose a competitive threat to the American aircraft industry. Eads did not have to remind his listeners that the British project had ultimately produced the notorious Comets, which not only were much more expensive to operate than contemporary propeller-driven aircraft, but also had the unfortunate habit of falling apart en route. (The midair explosion of the prototype Soviet SST at the 1973 Paris Air Show may presage a similar future for SSTs.)

Summary

When the SST program was launched in 1963, the nation, in reaction to Soviet space successes and a supposed strategic missile gap, was much concerned with re-establishing the supremacy of American technology. An enormous program had been embarked upon designed to ensure that the first man on the moon would be an American. A tremendous buildup of American offensive strategic missiles was in process. Any area in which American science and technology was not the undisputed world leader was considered a potential source of threats to the national security.

Despite this technological hysteria, the SST project was not initiated blindly. Partly as a result of the SST's long gestation period, many of the economic and environmental constraints on the aircraft were clarified during the initial stages of the program. But these constraints—on aircraft size, sonic-boom intensity, engine noise, and performance characteristics—were then largely ignored when technological difficulties arose. And unanticipated data—low public tolerance of sonic boom and engine noise, possible serious impact on the stratosphere—were accepted grudgingly, if at all; SST proponents tended to regard these unforeseen problems as inevitable, imaginary, or avoidable through additional research.

By the late 1960s, the overriding concern with the national security had receded. The Vietnam War had taught its bitter lessons about governmental limitations and fallibility. It became possible to question how strongly the international position of the United States depended upon the SST project and to raise the issues of its environmental impact and economic viability. The Nixon administration thus received a fresh opportunity to conduct an assessment of the costs and benefits of the SST program and to act on what they found.

In retrospect it appears that, by the time this opportunity arose, virtually all those in the administration and in Congress with direct responsibility for reviewing the program were unwilling to contemplate seriously the possibility of its termination. An informal alliance had formed to protect the SST project, with key members being the Federal Aviation Administration, the Boeing Company, and the Senators from Washington State, Boeing's home.

Not only was the public interest excluded, but considerable efforts were made to keep adverse information from the public and to soothe it with deceptive statements when important objections were raised by outside experts. Attempts, often largely successful, were made to suppress unfavorable reports on the program—and, when these attempts failed, to commission other studies which would criticize or "supercede" them. The public could not even depend upon the government to enforce the terms of the SST contract with Boeing.

The SST issue was ultimately "taken to the public" after governmental officials and agencies had repeatedly proven their unwillingness to act in the public interest. It is difficult not to be impressed by the effectiveness of the small number of Congressmen and scientists who dedicated a substantial part of their energies to criticizing the SST and leading the fight against it. Both "insider" and "outsider" scientists made indispensable contributions. Richard Garwin and Laurence Moss had acquired some of their expertise through service to the executive branch, a fact that helped bring attention to Garwin's views. But William Shurcliff, who was perhaps the most effective of the scientists who campaigned against the SST, informed himself about the project using only public information and fought against it entirely in his spare time.

In the public debate over the SST, the project's proponents tried to sell the *idea* of a supersonic passenger plane, emphasizing the possible dire consequences for the American aeronautics industry and the American economy if the SST were to be abandoned. Its critics, on the other hand, attacked the deficiencies of the *actual* aircraft whose construction Congress was being asked to fund and cited the possible disastrous environmental impact of the SST. This contrast—the promise of ideal technology vs. defects in actual designs—has become a common theme in debates on the exploitation of new technologies, as has the raising of the specters of foreign competition and of environmental disaster. If technological decisions are to be made responsibly, however, glib generalizations must be avoided and the proposed project evaluated on its merits. The Congressional hearings and debates on the SST provided the opportunity for such an evaluation.

Although Senator Proxmire and a small number of his fellow Congressional critics had opposed the SST since the project's inception, Congress as a whole did not seriously reconsider the SST until 1970. Indeed, until the antiballistic missile debate of 1969, Congress had never really challenged the administration on a major high-technology program. The SST issue was seriously examined by Congress only after it had become the subject of a full-scale national debate, led by environmental groups and largely informed by independent scientists. With the environment suddenly a major national issue and with the economy in poor health and the budget tight, it became increasingly difficult for Congress to accord high priority to annual contributions of hundreds of millions of dollars to the SST. Although the SST first attracted national attention because of its adverse environmental impact—noise, sonic boom, stratospheric pollution—in the end Congressional support was withdrawn mainly on priorities grounds. Expert testimony that the aircraft was technologically premature and economically marginal clinched the case against it.

The public battle over the SST was a landmark in the history of government and technology. It demonstrated that public opinion can, on occasion, play a crucial role in setting limits on what is acceptable public policy for technology. It established that the side effects of some technologies can be so serious that it may be better to leave those technologies unexploited. The SST fight also showed just how little scrutiny government programs receive from individuals who are in a position to learn the facts and are free to speak publicly. If the SST project had been subjected to serious, sustained, independent evaluation at an earlier stage, it might have been easier for the government to modify or terminate it. It seems quite possible to us that, given competent critics whose voices both the public and the government could hear, the SST program might well have been canceled sometime during the period 1964-1968. Studies that were done in 1963 had been forgotten three years later. It was during this period that it became clear how objectionable the SST's sonic booms would be to the public and how far short the SST design would fall of the initial program objectives. The fact that an administration official could state in 1966 that "we're all-out for economics now and to hell with the boom" dramatizes how insulated the government can become from social and technical realities.

NOTES

1. This is an excerpt from a guest editorial by Senator Barry Goldwater (R-Ariz.) in the *New York Times,* December 16, 1970, p. 47. The editorial was based on a Senate speech by Goldwater [*Congressional Record* 116 (1970): 34766-69], and the speech was based in turn on a Library of Congress report "SST: Issues of Environmental Compatibility," by George N. Chatham [reprinted in the *Congressional Record* 116 (1970): 39749-58]. Chatham's report is marred by obvious bias and by a number of technical inaccuracies. Several of the "far-fetched" arguments against which Goldwater rails appear to have been invented by Chatham in a section of his report entitled "Scenarios of Doom."

2. The first hearings were held in 1960 [U.S. Congress, House, Committee on Science and Astronautics, *Supersonic Air Transports,* 86th Cong., 2nd sess., May 17-24, 1960]. They were intended chiefly to promote the SST. Nevertheless, a number of witnesses gave candid assessments of technical difficulties that the SST would face.

3. President Kennedy's message to Congress on the SST is reprinted in U.S. Congress, Senate, Committee on Appropriations, *Independent Offices Appropriations, 1964: Supersonic Transport Development Program,* 88th Cong., 1st sess., June 19, 1963, pp. 1978 ff.

4. Ibid., pp. 2002, 2031.

5. Ibid., p. 1994.

6. Ibid.

7. "Final Report: An Economic Analysis of the Supersonic Transport," Stanford Research Institute Project no. ISU-4266, p. II-1, quoted in Don Dwiggins, *The SST: Here It Comes Ready or Not,* (Garden City, N.Y.: Doubleday, 1968), p. 15. The Dwiggins book is a good general reference on the SST through 1967.

8. *Public Papers of the Presidents of the United States: John F. Kennedy 1963* (Washington, D.C.: Government Printing Office, 1964), p. 441. Reprinted in the *New York Times,* June 6, 1963, p. 25.

9. President Kennedy's SST message, Ref. 3, pp. 1982, 1987f.

10. Some good references on the confusing period of SST design competition and detailed design are Dwiggins, *The SST;* John Mecklin, "$4 Billion Machine that Reshapes Geography," *Fortune,* February 1967, p. 113; Charles J. V. Murphy, "Boeing's Ordeal with the SST," *Fortune,* October 1968, p. 128; H. L. Nieburg, *In the Name of Science* (Chicago: Quadrangle, 1970), pp. 324-333; and William A. Shurcliff, *SST/Sonic Boom Handbook* (New York: Ballantine, 1970). An excellent survey of the entire SST history is Ian Douglas Clark, *Technical Advice and Government Action: The Use of Scientific Argument in the Ten-Year SST Controversy* (unpublished manuscript, dated June 1972, prepared in part for the Science and Public Policy Seminar at the John Fitzgerald Kennedy School of Government, Harvard University).

11. Quoted in the *New York Times,* October 31, 1966, p. 1.

12. U.S., National Aeronautics and Space Administration, *The Supersonic Transport, A Technical Summary*, NASA Technical Note no. D-423, June 1960, p. 21.

13. On neglect of noise in the engine design competition, see Michael Wollan, "Controlling the Potential Hazards of Government-Sponsored Technology," *George Washington Law Review* 36 (1968): 1125. Incidentally, it seems likely that if the SST program had been restarted in 1971, new engines would have been designed, probably of the quieter turbofan type; see *New York Times,* May 15, 1971, p. 60.

14. Shurcliff's *SST/Sonic Boom Handbook* summarizes the tests of boom acceptance and gives references; it also contains a good introduction to the causes and effects of sonic boom. The FAA's side of the story is told by John O. Powers and Kenneth Powers, *The Supersonic Transport—The Sonic Boom and You,* FAA Pamphlet, September 1968.

15. Technical advisory reports suggesting the unacceptability of SST sonic boom include one by the Office of Science and Technology, August 1967, and a three-part report by the National Academy of Sciences (NAS)-National Research Committee on SST-Sonic Boom. The former is discussed in the *New York Times,* August 8, 1967, p. 1; the latter is summarized in the *NAS News Report* 18, nos. 1, 3, and 6 (1968). Deficiencies in the NAS report are discussed in Chapter 4.

16. Quoted in Donald F. Anthrop, "Environmental Noise Pollution: A New Threat to Sanity," *Bulletin of the Atomic Scientists,* May 1969, p. 15. The contention that people accommodate themselves to the sonic boom is critically discussed in Karl D. Kryter, "Sonic Booms from Supersonic Transports," *Science* 163 (1969): 359.

17. B. K. O. Lundberg, "Supersonic Adventure," *Bulletin of the Atomic Scientists,* February 1965, pp. 29-33. Lundberg, director general (now emeritus) of the Aeronautical Research Institute of Sweden, has written and spoken extensively against the SST since 1960. (For references see, e.g., Shurcliff, *Handbook*, pp. 149-150.) It was he who developed many of the technical arguments used by Shurcliff and other SST opponents.

18. The ad hoc SST Advisory Committee report and related documents are reprinted in the *Congressional Record* 115 (1969): 32599-32613, and are partially reprinted as an appendix to Shurcliff, *Handbook*. The Garwin Report is reprinted in *Congressional Record* 117 (1971): 32126-29. In addition to these comprehensive assessments, analyses of special aspects which were more favorable to the SST were also received from the airlines [*Congressional Record* 115 (1969): 32611-13] and aeronautical experts.

19. *Congressional Record* 115 (1969): 32600.

20. Ibid.

21. Ibid., p. 32601.

22. Ibid., p. 32602.

23. Ibid., p. 32603.

24. *Congressional Record* 117 (1971): 32129.

25. President Nixon's speech is reported in the *New York Times,* September 24, 1969, p. 1.

26. Ibid.

27. For example, President Nixon said, in a televised conversation with a group of

schoolchildren: "In 1980 you will go from Washington to Los Angeles in an hour and a half. That is how fast we'll be moving with the new planes that will be available then." (Quoted in the *New York Times,* August 26, 1970, p. 26).

28. Various Senate votes on the SST are tabulated in Harry Lenhart, Jr., "SST foes confident of votes to clip program's wings again before spring," *National Journal,* January 8, 1971, pp. 43-58. This useful article contains a detailed discussion of the 1969 Senate decision on the SST.

29. The ultimately successful efforts by congressmen and environmentalists to force the release of the ad hoc committee report and the Garwin Report are described in Chapter 4. The discussion of the ad hoc committee report before the Senate Appropriations Committee, November 1969, is notable for the Department of Transportation's strenuous attempt to rebut and denigrate it. The rebuttal appears in U.S. Congress, Senate, Committee on Appropriations, *Department of Transportation Appropriations for Fiscal Year 1970,* 91st Cong., 1st sess., pp. 663-783; see also pp. 586-587, 788.

30. Quoted in George Lardner, Jr., "Supersonic Scandal," *New Republic,* March 16, 1968, p. 17.

31. Quoted in "SST foes confident," Ref. 28, which also mentions some of the arm-twisting in question. For more, see the *New York Times,* November 30, 1970, p. 81, and December 4, 1970, p. 1.

32. "SST foes confident," p. 48.

33. This comparison appears in Department of Transportation testimony at several hearings. It was used in Chatham's report, and also by J. J. Harford, executive secretary of the American Institute of Aeronautics and Astronautics and a leading pro-SST spokesman, in a guest column in the *New York Times,* June 13, 1970, p. 30.

34. One DOT balance-of-payments estimate was a projected \$72 billion loss through 1990 if the SST were not developed; reported in the *New York Times,* August 28, 1970, p. 1. The adverse balance-of-payments effects of the U.S. SST, not taken into account in this estimate, included the increase in international air travel that the SST would need to stimulate in order to create a demand for a large enough number of SSTs to justify the industrial investment. More of these passengers would be Americans flying on foreign-owned SSTs than foreigners flying on American-owned SSTs, and these Americans would furthermore spend additional dollars abroad. On this basis, the balance-of-payments panel of President Nixon's SST Review Committee concluded that the overall balance-of-payments effect of an American SST would be *negative* [*Congressional Record* 115 (1969): 32599].

35. Advertisement by American Labor and Industry for the SST in the *New York Times,* March 8, 1971, p. 22.

36. U.S. Congress, Senate, Committee on Appropriations, *Department of Transportation and Related Agencies Appropriations, Fiscal Year 1971, Part 2,* August 27, 1970, p. 1336. Shaffer's statement that sonic booms are not destructive was made in a speech to an industrial group November 17, 1969, and repeated to the Senate [U.S. Congress, Senate, Committee on Appropriations, *Department of Transportation Appropriations for Fiscal Year 1970,* 91st Cong., 1st sess., November 25, 1969, p. 591].

37. Senator Proxmire's letters to the committee members and their replies are reprinted in U.S. Congress, Joint Economic Committee, *Economic Analysis and the Efficiency of Government, Part 4-Supersonic Transport Development,* 91st Cong., 2nd sess., May 1970, pp. 1017-1030. A final round of letters on the SST from various government departments in August 1970 are reprinted in U.S. Congress, Senate, Committee on Appropriations, *Department of Transportation Appropriations,* 91st Cong., 2nd sess., August 27, 1970, pp. 1291-1328. This time the letters were solicited by the White House itself, and the administration's back was to the wall as a result of the unexpectedly close House vote in May. As might be expected, these letters were SST endorsements. (See pp. 1601-1602 of these same hearings.)

38. *Congressional Record* 116 (1970): 32609.

39. *Economic Analysis* (Ref. 37), pp. 1029f.

40. Representative Yates recounted his conversation with DuBridge in U.S. Congress, House, Committee on Appropriations, *Civil Supersonic Aircraft Development (SST). Continuing Appropriations, Fiscal Year 1971,* 92nd Cong., 1st sess., March 1, 1971, p. 41.

41. Reprinted in the *Congressional Record* 116 (1970): 41594.

42. *Economic Analysis,* pp. 999f.

43. The stratospheric-contamination question was reviewed by two successive groups of scientists in MIT-organized summer workshops in 1970 (Study of Critical Environmental Problems—SCEP) and in 1971 (Study of Man's Impact on Climate—SMIC). The SCEP report was largely in agreement with Train's testimony and expressed "a feeling of genuine concern" about possible SST effects. See *Congressional Record* 116 (1970): 27382; Senate Department of Transportation Appropriations Hearings, August 27, 1970, pp. 1588-1590; and SCEP, *Man's Impact on the Global Environment* (Cambridge: MIT Press, 1970), pp. 15-18, 64-74. The SMIC report, *Inadvertent Climate Modification* (Cambridge: MIT Press, 1971), reexamined the issue. In 1971, a University of California (Berkeley) chemist, Harold Johnston, pointed out that nitrogen oxides from the SST exhausts might have a much more destructive impact on the stratospheric ozone (*Science* 173 (1971): 517). The SMIC group does not appear to have been equipped to give this suggestion a serious review, and the matter will probably not be settled until much more sophisticated models of stratospheric dynamics are developed.

44. *Economic Analysis,* p. 913.

45. Ibid., p. 909. Garwin's comparison between the SST and jumbo jet takeoff noise was unsuccessfully disputed by the Department of Transportation: the technical witnesses agreed with Garwin. See the Senate Department of Transportation Appropriation Hearings, August 1970, p. 1492. See also the *New York Times,* May 23, 1970, p. 44, and May 29, 1970, p. 28.

46. For example, according to Peter Koff, vice-president of the New England Sierra Club and assistant director of the Citizens League Against the Sonic Boom, the most influential factor in New Hampshire Senator Robert McIntyre's decision to vote against the SST was his ninety-minute conversation with Garwin.

47. John Lear discusses the history of the Coalition Against the SST in *Saturday Review,* January 3, 1971, pp. 63 ff. See also "SST foes confident" (Ref. 28). Moss's activities are reported further in *Technology Review,* January 1971, pp. 72-73.

48. The statements were inserted into the *Congressional Record* 116 (1970): 31674-31678, of September 15, 1970, by Senator Fulbright and also received good press coverage. Only Henry Wallich, of the sixteen economists contacted, did not favor termination of government support for the SST. Wallich, then chief consultant to Treasury Secretary Kennedy, said that "a strong case could be made" for banning SSTs, but that if the Concorde is viable, then the United States should proceed with the SST. However, he added: "A case can be made for postponing immediate construction of the plane in favor of continued development work. This may lose us some part of the market for the plane, but we might compensate by having a better plane later at lower cost to the government."

49. U.S. Congress, Senate, *Department of Transportation Appropriations Hearings,* August 1970, p. 1718.

50. Ibid., p. 1719.

51. Ibid., pp. 1736-1739.

PART II

Advising or Legitimizing?

It is true that Government is getting a great deal of
advice, and some information, from the legions of
advisory bodies which it creates. I am much less clear
on what happens to the advice or who is listening. I
do know that very little of the advice from most
Presidential advisory bodies ever seeps through to the
President himself. Most of it is lost through evapora-
tion, some leaks out on staff advisors to the President, and no one can say with certainty how much of
it feeds into policy decisions. . . .

In my experience, nothing was simpler than to set
up an advisory group. It started wheels turning, it
bought time, it was a surrogate for action, and it
produced a kind of structural grandeur. It implied
that someone was taking charge of the problem, and
perhaps that things would work out. This is the way
of governments.[1]

—William D. Carey, former
assistant director of the
Bureau of the Budget, in Congressional
testimony.

INTRODUCTION

Confidentiality

THE existence of the national science advisory establishment is hardly a secret. Indeed, administration spokesmen never tire of reminding the public that their agencies have consulted with eminent experts on every issue a citizen could conceivably worry about. Whenever the public appears insufficiently impressed, a new advisory committee is appointed. The tremendous quantity of this advice and the quality of the advisors is even more visible to the scientific community, where virtually all scientists are aware of established colleagues, some of them quite eminent, who serve as government advisors. Very seldom, however, are outsiders able to catch a glimpse of the advice itself or to see how it relates to the actions taken by the government (i.e., executive-branch) officials who are its recipients.

By strongly enforced custom, the relationship between a science advisor and governmental advisee is confidential. The purpose of this confidentiality is to leave the government official free to accept or reject advice without political embarrassment. This is nice for the official, but it often leaves Congress and the public in the dark. Advisory reports prepared for administration officials are often the only authoritative assessments of relevant technical issues. If these are not available to Congress and the public, then policy making for technology is not subject to ordinary democratic safeguards. In fact, even the internal executive-branch decision-making process suffers because the same confidentiality which keeps important technical information from the public also makes it less accessible to other parts of the executive branch—even the President.

Advisory confidentiality often extends far beyond the requirements of military security or even the legal limits on governmental reticence specified in the Freedom of Information Act.[2] Consequently political and legal pressure can, if exerted with enough persistence, eventually result in the release of important documents. Or the documents can be "leaked." Such documents form the basis for the case studies presented in Chapters 4 through 7: the SST (again), the antiballistic missile (ABM) debate, the safety of defoliants, and the banning of cyclamates from food. In each of these cases, advice was disregarded and advisory reports were suppressed. The confidentiality of the advisory system allowed it to be used not merely to inform government officials but also to mislead Congress and the public.

33

Advising or Legitimizing?

Government officials are often under strong pressure to defend political and bureaucratic interests. They receive almost instantaneous feedback when their decisions appear to conflict with these interests, while the public interest has few spokesmen who can reach them even indirectly. The Watergate scandal has made it perfectly clear that government officials may sometimes find it expedient to respond to motivations other than the public interest. It is not surprising that officials should wish to use the advisory apparatus to hide their true motivations and give what are actually political decisions the appearance of being technical—i.e., apolitical.

In spite of such propaganda, final decisions on these matters are of necessity always political. Science advisors can help to estimate the costs and benefits of proposed courses of action, but such analyses are usually not decisive. (In this respect the comprehensive SST reviews commissioned by President Nixon in early 1969, in which the "costs" overwhelmingly outweighed the benefits, were atypical.) The political process must determine the relative weight which is to be accorded to each "cost" and each benefit. There is no other way in which, for example, the time saved for the passengers in an airplane may be compared to the annoyance caused to those living near airports by the engine noise. In a democracy the political process should reflect as accurately as possible the informed preferences of the people who are going to have to live with the decisions. Insulating government officials from public accountability behind a shield of silent "experts" does not place policy above politics. It simply subjects it to the narrower political considerations that prevail within the administration.

If an administration spokesman wishes the outside world to believe that a policy was adopted for technical rather than political reasons, the fact that his agency has consulted some of the most eminent experts will tend to persuade Congress, the public, and even other government agencies to accept the policy. The mere existence of the advisory system can be used for this "legitimizing" function even when the decisions being defended fly in the face of the information and analyses the advisors have provided—as long as the advice itself is kept confidential.

The most frequent means by which the public is misled is through the incomplete statement. Typically, an administration spokesman says that his agency, after consulting the greatest authorities, has decided to do X. The spokesman neglects to mention, however, that the experts have given mostly reasons why X might be a bad idea. Concerned citizens cannot check what the experts actually said, because their reports are kept secret.

The case studies in Chapters 4 through 7 illustrate the spectrum of other devices by which the federal executive branch's science advisory system has been abused:

• • • Officials can selectively make public only advisory committee reports that present *positive* terms in a cost-benefit calculation. This happened in connection with the noise suppressors for the SST. The government neglected to mention that their weight would be comparable to the total payload of the aircraft.

· · · Advisory committee reports that are made public may be so written that they are seriously misleading, at least to the press; and political and institutional pressures may prevent the issuance of a proper clarification. This happened with a National Academy of Sciences report on damage from SST sonic booms. (Both of these cases are discussed in Chapter 4.)

· · · Sometimes government spokesmen don't content themselves with half-truths; they simply lie. William Magruder's statement (quoted in the preceding chapter) that the scientific authorities who counseled the government had agreed that the SST will have no significant adverse environmental effects was a lie. Similar misrepresentations occurred in public statements about the effectiveness of the Safeguard ABM system. (This case is discussed in Chapter 5.)

· · · When the government has exclusive access to certain information about a public health hazard, it can simply suppress it instead of acting on it. This occurred in connection with the defoliant 2,4,5-T, which causes birth defects (Chapter 6).

· · · An advisory committee can be manipulated so that it adapts its advice to the political needs of the official being advised. This happened in the controversy that erupted after it was learned that cyclamates can cause cancer, when officials tried to allow cyclamate-sweetened foods to be sold as "nonprescription drugs" (Chapter 7).

These have been more controversial than most technical issues, but they do not appear to be unrepresentative of the manner in which the executive branch treats technical advice when major government or industrial interests are at stake. But even if these cases were unrepresentative, they would still be of sufficient importance in themselves to justify a reappraisal of the advisory system.

Preventing Advisory Abuses

In September 1972 Congress passed the Federal Advisory Committee Act, after several years of Congressional hearings on advisory committee abuses. The new law represented Congress's recognition of the importance of ensuring that advisory committees are not used to undermine the processes of democratic government. The net effect of this act is mainly to make explicit the applicability to advisory committees of the Freedom of Information Act. In accordance with the new law, many advisory committees that formerly met behind closed doors are now announcing their meetings and conducting at least part of their business in public. But unfortunately the same exemptions which have weakened the Freedom of Information Act also apply to the Advisory Committee Act. And the largest science advisory organization of all—the

National Academy of Sciences—appears to be entirely exempt from this law, along with all other advisory agencies that work on contract. (The impact of the Advisory Committee Act is discussed in more detail in Chapter 9).

Quite apart from the new law, individual science advisors have occasionally been willing to bring to public attention important information and analyses that they felt were being disregarded in the executive branch. For example, as we mentioned in Chapter 2, Richard Garwin acceded to several requests to appear before Congressional committees and give his views on the SST despite the fact that he had chaired President Nixon's technical advisory committee on the SST. (Most Presidential advisors in Garwin's position would have claimed Executive privilege and refused to testify.) A year earlier, in 1969, Garwin had taken it upon himself to write to every member of the Senate, and to meet privately with many Senators, in order to explain to them the technical and strategic defects of the proposed ABM system. During this same period, Garwin was a member of the elite President's Science Advisory Committee (PSAC), having previously served on PSAC under Presidents Kennedy and Johnson and been reappointed to a second term by President Nixon. Despite the fact that all PSAC members except the chairman (the President's science advisor) were prominent non-governmental scientists who devoted only a few days each month to their advisory duties, PSAC was expected to support the President and Dr. Garwin's lack of "loyalty" to the Nixon administration reportedly angered key White House officials. They were presumably further displeased when the White House was forced by a suit under the Freedom of Information Act to release the long-suppressed report of Garwin's SST advisory panel, which had recommended termination of Boeing's contract.

Abolish Science Advisors?

Frustration with science advisors like Dr. Garwin was partially responsible for the Nixon administration's decision in early 1973 to abolish the entire White House science advisory apparatus—PSAC, the Office of Science and Technology, even the position of Presidential science advisor. (The numerous lower-level science advisory committees were not directly affected.) The official explanation for this change was that outside science advice was no longer needed at the Presidential level, but could instead be provided through the various federal departments and agencies—in particular, the National Science Foundation. This argument of course ignores one of the principal rationales for setting up the Presidential science advisory system in the first place: the President's need for technical analyses unbiased by bureaucratic self-interest.

Few scientists—even the most unreserved critics of the executive branch's science advisory establishment—greeted the news of PSAC's demise with rejoicing. The abuses of the advising system arise out of its political exploitation, and the White House appears to have abolished PSAC precisely because it was not exploitable enough. The political advantages expected from tactics such as public endorsement of the SST and ABM projects by President Nixon's science advisors never materialized—or were offset by the activities of more independent advisors like Garwin.

The advisory system has legitimate informational and political functions to perform, and it is likely that some sort of Presidential science advisory structure will eventually be reestablished. What can it realistically be expected to accomplish? The goals and limitations of science advising in the executive branch are the subject of the following chapter.

REFERENCES

1. U.S. Congress, House, Committee on Government Operations, *Presidential Advisory Committees, Part 1,* 91st Cong., 2nd sess., March 19, 1970, p. 160.

2. The Freedom of Information Act [5 *United States Code* 552], which was enacted in 1967 after eleven years of studies and Congressional hearings, requires that most government documents be available for public inspection. Nine categories of documents are excepted, including national security secrets, trade secrets, personnel files, and "inter-agency or intra-agency memorandums or letters which would not be available by law to a party other than an agency in litigation with the agency." Unfortunately, the Act specifies no penalties for delay, and executive branch agencies have frequently succeeded in withholding information from Congress and the public until it was no longer timely. We give several such examples in the chapters which follow.

CHAPTER 3

The Uses and Limitations of Science Advisors

The Need for Science Advice

Although the vast majority of government scientific advisors are concerned with relatively small decisions (such as the choice of materials to be used in military equipment) or with the technical review of grant requests from their fellow scientists, we focus in this book on the roles played by high-level science advisors in major policy decisions: whether to proceed with the SST development program, whether to ban most uses of cyclamates or DDT, whether to deploy the Safeguard antiballistic missile system. For such decisions the primary service which the advisors provide is not information—the decision maker usually has plenty of that supplied by his own technical staff and that of government contractors. The advisors' major contributions are analytical and critical.[1]

BYPASSING CHANNELS

An occupational disease of bureaucracy is self-deception. Power can be concentrated at the top of bureaucratic hierarchies; but information cannot be concentrated, only filtered. By the time it reaches the officials in charge, information generated within a bureaucracy will ordinarily pass through the hands of several lower-ranking bureaucrats each of whom has the power to delete but few of whom have anything worthwhile to add. Doctoring of reports to alter their conclusions is not unheard of.[2] Even with the best of intentions, a large bureaucracy intellectually insulates its higher officials. The people on top may have the authority to make choices, but the options from which they choose and the information on which they base their choice are prepared by their subordinates.

After a bureaucracy has been in existence for a few years, it will have made certain decisions, established certain operating procedures, and solidified certain relationships with other powerful institutions. All of these arrangements constrain the options and the information available within the bureaucracy. Thus are born bureaucratic procedures and bureaucratic truth.

Leading government officials are usually eventually forced to respond to nonbureaucratic perceptions of reality—by the newspapers, by Congress, or by the courts. But an astute leader will want to know in advance the likely responses to his actions, and he will not wish to be overly constrained by bureaucratic precedent. In order to obtain a candid response on these matters, he must obviously turn to people whose own positions are sufficiently secure and independent that they will not be much influenced by the reception their advice is accorded. Hence the need for outside advisors. This need is particularly acute in highly technical areas, where government officials often cannot entirely trust their own judgment and where the outside advisors may have a considerably broader expertise than regular government employees.

Besides helping to prevent the government from cutting itself off from reality, the science advisory system has sometimes also acted as an excellent conduit for new ideas and information—both within the government and between the government and the scientific community. This has been made possible partly because of the way science advising was organized and partly because of the nature of the scientists themselves. Committees advising different government departments on similar subjects are frequently intimately interconnected by overlapping memberships. The inner circle of the science advisory community—the few hundred scientists who are on everyone's list of the "right names"—see each other in numerous other capacities in their professional activities and as representatives of their universities or corporations. These scientists are in touch with developments in their parts of the scientific community and typically serve simultaneously at several levels in the advisory establishment. They are thus able to cultivate a flourishing grapevine, whose narrower runners are the telephone lines and whose main branches are the transcontinental jet routes—and whose roots are nourished by the larger scientific community. Good scientists know that they must always be open for new ideas, and they have learned from repeated experience that the important new ideas often arise outside the "establishment." As a result, the science advisory grapevine—and the larger, informal communications network of science of which it forms a part—can provide pathways for a rapid flow of ideas and information from the scientific community or from the lowest levels of the government directly to the highest officials, bypassing the slow and selective bureaucratic filter.

IDENTIFYING THE CHOICES

Perhaps the most difficult part of governmental decision making—just as in scientific research—is the recognition of the important problems. Since scientists are more familiar with the technical facts than are government officials, they are often the first to perceive such problems. For example, the 1960 NASA report quoted in Chapter 2 (see page 14) pointed out the importance of minimizing takeoff and landing noise in the design of the SST engines. Unfortunately this advice had been forgotten by the time the choice of SST engines had to be made. This example illustrates another moral: the need for continuous technical review of important programs. One of the most serious deficiencies in the system of ad hoc advisory panels and committees is that while committees come and go, the problems remain.

CONSIDERING POLICY IMPLICATIONS

Should science advisors answer only purely technical questions and seek merely to identify but not address issues requiring political choice? In practice, it has been found impossible to make such a clean separation between the functions of science advisor and policy maker. At the higher levels of government, science advisors have been repeatedly called upon to help make policy as well as render technical judgments.

One reason why the roles of advisor and decision maker cannot be clearly separated is that decisions on questions like the safety of a new drug or the environmental impact of the SST are never in practice based on adequate information. The various benefits and costs are usually largely a matter of guesswork. And postponing a decision until better information becomes available in itself constitutes a decision. Obviously, only a person familiar with the technical information is in a good position to estimate the risks arising from uncertainty. And an advisor who understands the technical issues may also be helpful in judging how heavily to weigh these issues against other, nontechnical considerations.

Because public officials must often rely upon the combined political and scientific judgment of their technical advisors, they tend to choose as advisors scientists whose political views are similar to their own. Presidential science advisors were routinely selected on this basis. But while shared assumptions may improve communication, they may also effectively result in political views determining technological policies without sufficient regard for technical considerations. In some cases balance has been achieved within the executive branch when opposing factions have established their own advisory groups, each having different political biases.[3] Thus, the President's Science Advisory Committee shared the interests of Presidents Eisenhower and Kennedy in a nuclear test-ban treaty and helped them stand up to the prophecies of doom which arose from Pentagon and Atomic Energy Commission experts whenever the prospects of negotiation with the Soviet Union appeared to brighten. The impossibility of avoiding some political bias in advisory groups is of course an additional reason why Congress and the executive branch should each have their own advisors—even if executive-branch advisory reports were to be made freely available.

A Success Story

An example of the operations of the science advisory system at its best will make some of the abstract discussion of the last several pages more concrete. It should also serve to counterbalance the more disillusioning stories that occupy the next four chapters. The example concerns a President's Science Advisory Committee (PSAC) report on the long-term hazards of pesticides.[4]

BACKGROUND

The insect-killing properties of dichloro-diphenyl-trichloroethane (DDT) were discovered in 1939 by the Swiss chemist Paul Müller. In the following years the chemical was found to kill an almost incredible number of insect and even rodent pests—ranging from malaria-bearing mosquitos, through the cotton bollworm and the spruce-budworm, to rats and bats. Public enthusiasm for the new chemical was almost unbounded, and in 1948 Müller was rewarded by a Nobel Prize for his discovery.

The popularity of DDT unleashed within the chemical industry a great search for other synthetic organic pesticides. By the mid-1960s many hundreds were being sold in the United States in tens of thousands of preparations with annual retail sales amounting to more than a billion dollars. This enormous market had been created with substantial help from the U.S. Department of Agriculture (USDA), which was by statute responsible for the promotion of agriculture as well as the regulation of pesticide use so as to protect the public health. (The Environmental Protection Agency was given the authority to regulate pesticide use in 1970.) County agricultural extension agents, who had substantially worked themselves out of a job as they successfully fostered the modernization of American agriculture, had joined the chemical company salesmen in efforts to convince farmers to make massive and almost exclusive use of synthetic pesticides against all sorts of real and sometimes imaginary pest threats to their crops. Local governments and individual homeowners followed suit by using pesticides in great quantities to kill mosquitos, elm bark beetles, roadside brush, and innumerable other unwanted infestations.

In 1962 Rachel Carson, a biologist and writer of popular nature books, published *Silent Spring.*[5] The book presented dramatically and with painstaking documentation the basis for her concern about the impact of pesticide usage on the environment and on human health. From *Silent Spring,* the public learned a particularly surprising and frightening fact: after DDT is widely dispersed in a spraying program, its chemical properties result in its being absorbed out of the environment into the bodies of animals and returned to man in astonishing quantities in the milk, eggs, meat, and fish he eats.[6]

The fact that DDT migrates in the air and water and lasts for years without significant decomposition (and hence is labeled "persistent") have made it one of the few truly long-lived and global pollutants. Thus it was clear to Miss Carson that, if exposure to DDT was found one day to be a serious hazard to human health, it might very well be too late to do anything about it. When *Silent Spring* was published, the typical American already had about a gram of DDT stored in his fat.

Although it was unclear what the long-term human consequences of this exposure would be, by 1962 it already appeared to be disastrous for a number of other animal species. In particular, there were then indications that a number of birds of prey and sea birds were becoming extinct because DDT was making it impossible for them to reproduce successfully. On a local level, of course, it had become a common occurrence for a bird population to be virtually wiped out by the immediate toxic effects of DDT after the spraying of an area, with the fish in the streams, lakes, and offshore waters of the watershed often suffering the same fate. Because of the pervasiveness and persistence of DDT, it quickly became the focus of the national debate triggered by *Silent Spring*.

THE RESPONSE OF THE SCIENTIFIC ESTABLISHMENT TO *SILENT SPRING*

Silent Spring was greeted by agricultural and chemical industry spokesmen with a storm of opprobrium: "misinformed," "distorted," "hoax," and "fanatic" were typical characterizations.[7] The reviews of Silent Spring read most widely in the scientific community were also less than enthusiastic. In *Chemical and Engineering News* (October 1, 1962), the news magazine of the American Chemical Society, the review by William Darby, member and past chairman of the Food Protection Committee of the National Academy of Sciences' National Research Council (NAS-NRC), was entitled "Silence Miss Carson." In *Science*, the journal of the American Association for the Advancement of Science, I. L. Baldwin was slightly more moderate: he suggested that Miss Carson lacked perspective, dismissing her concerns about possible long-term public health hazards by stating that "most scientists who are familiar with the field, including government workers charged with the responsibility of safeguarding the public health, feel that the danger of damage is slight."[8] He did not, however, explain how this "feeling" could be substantiated in the absence of tests of pesticide chemicals for carcinogenicity (potential for inducing cancer), mutagenicity (potential for inducing genetic defects), or teratogenicity (potential for causing birth defects)—tests that had been urged in *Silent Spring*. Baldwin went on to stress his view that the benefits obtained from man's use of pesticides far outweighed the costs.

Finally, for a "careful and judical review of all the evidence available"[9] Baldwin referred to reports of a "committee of outstanding scientists", established by the National Academy of Sciences' National Research Council (NAS-NRC) to study the influence of pesticides on human health (Darby's committee), and a companion committee (chaired by Baldwin himself) which had been established to deal with pesticides and wildlife. Any readers who troubled to obtain copies of the reports Baldwin cited must have been disappointed. The reports are brief, superficial, and undocumented. For example, the report of Baldwin's committee devotes only two pages to the subject of "Wildlife Losses due to Pest Control in Agriculture" although an estimated 3 billion pounds of pesticidal preparations were being used in agriculture annually. Not only is the discussion quite cursory, but it seems also to avoid the more serious questions relating to pesticide use, such as the problem of persistent pesticides such as DDT being concentrated in food chains and their role in the worldwide decline—possibly even extinction—of certain species of birds. In general one gathers from the report that avoidable damage to wildlife should be minimized, but that when the choice is between unavoidable damage to wildlife—no matter how great—and the cancellation or reduction of a pest control program, the wildlife must go. Baldwin's committee had functioned under the ground rule that nothing appear in any of the reports that did not have unanimous approval within the subcommittee concerned.[10] This rule, in combination with the fact that a number of the committee members had close ties with the Department of Agriculture and pesticide manufacturers and were convinced pesticide enthusiasts, goes far in explaining the apparent evasiveness of the reports.

THE 1963 PSAC REPORT ON PESTICIDES

Silent Spring first appeared as a series of articles in *The New Yorker* in June 1962. Richard Garwin, then serving his first four-year term on the President's Science Advisory Committee (PSAC), was greatly impressed by Rachel Carson's arguments. At the next monthly meeting of PSAC, he distributed copies of her *New Yorker* articles and vigorously urged that PSAC conduct an independent investigation. Such a study was initiated several months later by Presidential science advisor (and PSAC chairman) Jerome Wiesner, after President Kennedy expressed concern about pesticides.[11]

Following the usual PSAC custom, also common on other science advisory committees of broad scope, Wiesner appointed an ad hoc panel—the Panel on the Use of Pesticides—which was commissioned to prepare a report to be submitted to the President after review by the full committee. The panel included three members of PSAC, four members from university faculties, the director of the Connecticut Agricultural Experiment Station, and a conservationist from the Audubon Society.[12] They met several times during an eight-month period to deliberate and to be briefed by experts on pesticides. The people from the Department of Agriculture regarded pesticide use as an all-or-nothing proposition, according to one member of the panel,[13] and they refused to discuss the individual merits or drawbacks of specific pesticides. Chemical company scientists in their turn emphasized the safety of their pesticides and the high costs of pesticide development. Rachel Carson was also called as a consultant. During a session lasting nearly a day, she impressed the panel members as being much more moderate and sensible than the more dramatic passages of her book had led them to expect.[14]

The panel soon reached a consensus that differed rather sharply from the prevailing opinions on pesticides in government and industry. They recognized that even "safe" pesticides have serious potential costs that must always be weighed against their benefits. Continued exposure to small amounts of persistent pesticides like DDT and eldrin can be harmful over long periods of time to wildlife and perhaps also to man. Chronic toxicity and the potential for causing cancer, genetic damage, or birth defects are much more difficult to detect than acute poisoning, but no less serious. The panel concluded that studies of such chronic effects in laboratory animals were urgently needed.

The panel was also critical of the prodigal use of pesticides in government efforts at total eradication by chemical means of particular insect species like the gypsy moth or fire ant. They argued that "acceptance of a philosophy of control rather than eradication . . . acknowledges the realities of biology" and pointedly urged that "Federal programs should be models of correct practice."[15] These comments may have been prompted by an event that occurred during the PSAC panel's deliberations. One panel member recalls a "long hot session" where Agriculture Department spokesmen discussed their plans to spray Norfolk, Virginia, with the persistent pesticide dieldrin, a chemical considerably more toxic than its cousin DDT, in an attempt to eradicate the white fringed beetle. Despite the panel's vigorous objections, the spraying was carried out on schedule.

President Kennedy reportedly often asked about the progress of the PSAC Pesticide Panel and urged speed in getting out the report.[16] He evidently liked the report when he finally received it, for when he released it to the public on May 15, 1963, he noted that he had "already requested the responsible agencies to implement [its] recommendations."[17] The report recommended, among other things, that "the accretion of residues in the environment be controlled by orderly reduction in the use of persistent pesticides. . . . Elimination of the use of persistent toxic pesticides should be the goal."[18] The report concluded with a quiet tribute to Miss Carson:

> Public literature and the experiences of Panel members indicate that, until the publication of "Silent Spring" by Rachel Carson, people were generally unaware of the toxicity of pesticides.[19]

The PSAC report was greeted by the press as a powerful vindication of *Silent Spring.* The reaction to the report in the scientific community was more cautious but no less significant. Although scientific controversy over various issues raised by Miss Carson certainly did not cease, the level of the discussion was raised from denunciation and personal vilification to reasoned argument. It may be grandiloquent to claim that PSAC acted as a high court of science on the pesticides issue, but the importance of PSAC's leadership in this case is undeniable.

More generally, the PSAC pesticide report, together with the broader-scope PSAC report *Restoring the Quality of Our Environment,* written two years later, can be credited with initiating a shift in the federal government's policy on pesticide use away from the massive insect-eradication programs of the Plant Pest Control Division of the Agriculture Department in the 1950s and early 1960s. Another decade of effort was required by organizations like the Environmental Defense Fund—a combination public-interest law firm and scientist-activist group—before commercial misuse of persistent pesticides was curtailed. (These developments are traced further in Chapter 10. And Chapter 6 is concerned with the herbicide 2,4,5-T, whose ability to induce birth defects was detected in laboratory tests undertaken following the recommendations of the PSAC pesticide report.)

The PSAC pesticide report thus accomplished several useful functions. It gave the President sound advice on pesticide policy—advice that he was not receiving from the Department of Agriculture or other regular government channels; it played a leading role in helping the scientific community come to grips with the problems of persistent pesticides; and it served to reduce the resistance within the government against further useful steps.

PERSPECTIVES

The executive-branch science advisory system deserves great credit for achievements like the PSAC pesticide report. But it must be kept in mind that, as the report itself admits, it was Rachel Carson who first brought the dangers of pesticides to general attention. If *Silent Spring* had not inspired a high-level review of pesticide hazards, the government would probably have continued to rely on such uncritical advice as that of the NAS-NRC committees chaired by Baldwin and Darby. The advisory system rarely develops significant new issues, responding instead to the initiatives of others. As a distinguished National Academy of Sciences panel noted somewhat ruefully:

When Presidential Task Forces, private foundations, or groups like the President's Office of Science and Technology or the President's Science Advisory Committee become involved, . . . the usual reason is that a specific area of concern has already reached near-crisis proportions or has otherwise captured the imagination of particularly articulate individuals (Ralph Nader and Rachel Carson come immediately to mind) or of unusually influential groups. The result is often a report that duplicates other efforts, or overlooks important considerations, or comes too late to exert any significant influence on the underlying technology, or is without a recipient other than the public at large.[20]

Advisory committees cannot entirely escape the diseases of the government bureaucracies to which they are attached. Because the government officials being advised often do not have adequate time to understand the issues involved in technological disputes, there is strong pressure on advisory committee members to compromise their differences and present a united front. "On the whole the greatest occupational hazard of advisory committees is not conflict but platitudinous consensus," according to Harvey Brooks.[21] Henry Kissinger, while still a Harvard professor, expressed the limitations of advisory committees even more forcefully:

The ideal "committee man" does not make his associates uncomfortable. He does not operate with ideas too far outside of what is generally accepted. . . .

Committees are consumers and sometimes sterilizers of ideas, rarely creators of them.[22]

It seems that no amount of improvement in the official science advisory system can obviate the need for the participation of independent scientists in democratic policy making for technology.

Political Uses of Advisors

In the first place, the high prestige of the National Academy of Sciences gives its recommendations an intrinsic merit of their own. It helps to have them behind you. . . . [I] used them to protect myself against other bureaucrats and politicians. For instance, with their backing I could appear more confidently at congressional hearings or before the public and not be fearful of having some politician or scientist claim the Commerce Department was all wrong because we hadn't consulted the right people.[23]

—Myron Tribus,
former Assistant Secretary
of Commerce for Science
and Technology

We come finally to the subject of the next several chapters: the political uses of science advisors. It is evident from the PSAC pesticide report that science advisors can contribute usefully to the government policy-making process. However, science advisory committees have been popular with government officials for other reasons as well. Bureaucrats are constantly involved in struggles to expand and defend their empires. This is a dangerous business, and what bureaucrats perhaps most desire from their science advisors is protection: protection against surprise by new technological developments and protection of their policies against political attack.

Technological surprise can take the form of an unforeseen technological development or an overlooked argument that a political opponent can pull out of the hat and turn to his own advantage. Unlike law, diplomacy, economics, or other traditional pursuits of public policy, science and technology are subject to radical new departures. And even if these departures surprise most scientists, they are at least the first to know.

If a government official deals with technology, he had better have scientific advisors who keep him informed of any new technical developments that may affect his position. In the aftermath of the shock of *Sputnik,* with its surprising revelation of Soviet technological prowess, President Eisenhower showed that he had learned the importance of this function by creating the position of Presidential science advisor and reconstituting PSAC, formerly an advisory committee to the Office of Defense Mobilization, as his personal science advisory committee.

The second essential function of science advisors, from the point of view of insecure government bureaucrats, is to act as high priests whose ministrations during the preparation of a policy are supposed to render that policy immune from political attack. A common strategy is exemplified by William Magruder's invocation of "the considered opinion of the scientific authorities" in support of his assertion that the SST would be environmentally harmless. When this ploy eventually failed, Magruder reverted to another standard device: the appointment of new and more cooperative committees of experts to study the problem.

The next chapters will give more examples to illustrate the ways that the science advisory apparatus has been used as an excuse to delay decision or action, to backstop an official or provide him with a justification for reversing policy, and generally to legitimize government actions and intimidate Congress and the public.

NOTES

1. A sympathetic but accurate portrait of the science advisory system has been given by Harvey Brooks in his essay "The Scientific Advisor," in *Scientists and National Policy-Making,* Robert Gilpin and Christopher Wright ed. (New York: Columbia University Press, 1964), pp. 73-96; reprinted in Thomas E. Cronin and Sanford D. Greenberg, eds. *The Presidential Advisory System* (New York: Harper & Row, 1969), pp. 40-57. Other essays in these volumes are also useful, and the standard literature on the science advisory system can be traced from their references. For a detailed discussion of the history and organization of the higher levels of the executive-branch science advisory system, see Frank von Hippel and Joel Primack, *The Politics of Technology: Activities and Responsibilities of Scientists in the Direction of Technology* (Stanford, Calif.: Stanford Workshops on Political and Social Issues, Stanford University, 1970) and references therein. See also U.S. Congress, House, Committee on Government Operations, *The Office of Science and Technology,* 90th Cong., 1st sess., March 1967.

2. Such was the fate of a report by Dr. Marvin Legator, chief of cell biology research at the Food and Drug Administration, to the FDA commissioner. See James S. Turner, *The Chemical Feast* (New York: Grossman Publishers, 1970), pp. 13-14. See also Chapter 7, below.

3. Anne H. Cahn showed in *Eggheads and Warheads: Scientists and the ABM* (Cambridge: Center for International Studies, MIT, 1971) that, with very few exceptions, the only Presidential science advisors on antiballistic missiles who *supported* ABM deployment were those who also served as Defense Department science advisors, and the only members of Pentagon science advisory panels who *opposed* the ABM were those who simultaneously served as Presidential advisors. Cahn furthermore showed that, if ABM advisors and activists were divided into pro- and anti-ABM groups, the groups differed strikingly in general political world-view. Policy and politics are hard to separate!

4. In his 1963 testimony before the Senate Appropriations Committee in support of continued funding for the Office of Science and Technology (through which PSAC was funded), Presidential Science Advisor Jerome Wiesner singled out the report discussed here—the PSAC report *The Use of Pesticides*—as exemplifying the way his office carried out its responsibilities. [U.S. Congress, Senate, Committee on Appropriations, *Independent Offices Appropriations for 1964,* Part 1, 88th Cong., 1st sess., 1963, p. 527.] Most PSAC reports dealt with military matters and are still secret.

5. Rachel Carson, *Silent Spring* (Boston: Houghton Mifflin Co., 1962); also available in paperback (New York: Fawcett World); first published in part in *The New Yorker,* June 16, 23, 30, 1962.

6. Because DDT has a very low solubility in water and a relatively high solubility in fat, it tends to concentrate in the fatty tissues of animals and in animal products with high fat content. The concentration in some fish and fish-eating birds, for example, has often been found to be many thousand times that in the body of water which supplied them their food. Other pesticides in the family of chlorinated hydrocarbons have the same property.

7. Quoted in Frank Graham, Jr., *Since Silent Spring* (New York: Fawcett World, 1970), p. 83.

8. I. L. Baldwin, *Science* 137 (1962): 1042.

9. Ibid.

10. Graham, *Since Silent Spring,* p. 52.

11. The pesticides inquiry was begun by a panel of federal officials but the responsibility was then shifted, apparently as a result of President Kennedy's concern, to PSAC. See Graham, *Since Silent Spring,* p. 61, and *Chemical and Engineering News,* May 27, 1963, p. 102.

12. Ibid.

13. The panel member quoted was William H. Drury, Jr., director of the Hatheway School of Conservation, Massachusetts Audubon Society. Quoted by Graham, *Since Silent Spring,* p. 83.

14. Interview with panel member Paul M. Doty.

15. U.S., President's Science Advisory Committee, *Use of Pesticides* (Washington, D.C.: The White House, May 15, 1963), pp. 18-19. Reprinted as "Report on the Use of Pesticides," *Chemical and Engineering News,* May 27, 1963, pp. 102-115.

16. Graham, *Since Silent Spring,* p. 83.

17. *Use of Pesticides,* p. iii.

18. Ibid., p. 20.

19. Ibid., p. 23.

20. U.S. Congress, House, Committee on Science and Astronautics, *Technology: Processes of Assessment and Choice* (Washington, D.C.: Government Printing Office, July 1969), p. 28. This report was prepared under the auspices of the NAS Committee on Science and Public Policy (COSPUP). Both COSPUP and the Technology Assessment panel were chaired by Harvey Brooks, Dean of Engineering and Applied Physics at Harvard. Brooks has been a consistently influential science advisor for many years.

21. "The Scientific Advisor," in *The Presidential Advisory System,* p. 52.

22. Henry A. Kissinger, *The Necessity of Choice* (New York: Harper and Row, 1961). The quotations are taken from the final chapter, entitled "The Policymaker and the Intellectual," p. 345.

23. Quoted in Claude E. Barfield, "National Academy of Sciences Tackles Sensitive Policy Questions," *National Journal,* January 30, 1971, p. 101.

CHAPTER 4

Not the Whole Truth:
The Advisory Reports on
the Supersonic Transport

One of the ways in which administration officials often mislead the public about the basis for their decisions is by releasing primarily (or exclusively) the information and analyses which support the administration position. The information so provided may be accurate, but it often is also totally misleading as to the true balance of costs and benefits. The long debate over the SST development project provides a number of examples of the selective release of information. Comprehensive advisory reports on the project's benefits and disadvantages were suppressed while the media were supplied with other reports which gave a misleading impression that certain objections which had been raised to the SST were not so serious after all.

The Comprehensive Reviews

In our discussion of the SST program in Chapter 2, we noted that immediately after taking office, President Nixon commissioned two high-level, comprehensive reviews of the SST program. One review committee was made up of senior officials from the relevant government departments and agencies, along with a representative of NASA, a member of the Council of Economic Advisors, and the President's science advisor. This committee reported to the President through the Secretary of Transportation, whose Department had primary responsibility for the project. Its charge was to consider whether continued federal funding of the SST development program was in the national interest.

The other comprehensive review which President Nixon commissioned was conducted by a panel of independent technical experts. This panel, which reported to the President through his science advisor, was chaired by Richard Garwin, an IBM physicist and a member of the President's Science Advisory Committee.

As we mentioned in Chapter 2, the panel reports of the interdepartmental SST Review Committee rejected the basic arguments which had been used to justify the SST project.[1] Furthermore, the panel considering the environmental impact of the SST concluded that the sonic boom would be intolerable if the plane were allowed to fly supersonically over populated areas and that SST airport noise would be a very serious problem.

The report of Garwin's panel was, if possible, potentially even more damaging to the project. In addition to the concerns raised by the interdepartmental review, Garwin's panel examined the extent to which the terms of the SST development contract with Boeing had been met. These terms had required Boeing to submit, by January 15, 1969, "a completely integrated design, fully substantiated by physical tests and detailed engineering analyses, . . . for a safe and economically profitable production version of the SST."[2] Garwin's panel observed, however, that

there are substantial grounds to believe that the Government could terminate the contract "for default." These grounds are of three types:

1. The fixed-sweep prototype, as proposed, will have take-off and landing runs some 50% longer, take-off and landing speeds very substantially higher, and other characteristics deficient with respect to the prototype required under the contract.

2. . . . The design is not fully substantiated as required by the contract.

3. It may be judged that the contractor has not demonstrated that the production airplane which follows from the prototype will be a "safe, economical . . . " commercial supersonic transport.[3]

These points were followed by a series of technical criticisms. The panel pointed out several aspects of the SST program of

high risk—among them the noise specifications. . . . More important and more fundamental is the fact that the estimated design payload constitutes only 7% of the aircraft gross weight, as contrasted with a realized 12-30% for a subsonic commercial transport of longer range. Our accuracy of design of structure, and our ability to calculate fuel consumption and adequate fuel reserves is not such as to insure that the payload will exceed 2%, which would have disastrous effects on the economics of the aircraft.[4]

The panel also found that, even if technical problems were overcome, the market might amount to only half of the Federal Aviation Agency's "conservative" estimate of 500 airplanes—a market too low to allow the government to recoup its investment in the development of the SST. Furthermore, Garwin's panel found it unlikely, because of the economic risks involved, that Boeing could obtain the several billion dollars of nongovernment financing required by the contract for the production phase of the program. The panel report observed dryly: "Both the government and the private sector can do much better with their money in other programs."[5]

After this devastating critique, the panel's primary recommendation for "termination of the development contracts and the withdrawal of Government support from the SST prototype program"[6] should have come as no surprise.

The interdepartmental SST Review Committee report and the Garwin Report are apparently the only comprehensive studies of the SST that President Nixon commissioned. Yet despite the strong negative recommendations of both of these reports, Nixon gave his go-ahead to the SST program in September 1969. As far as the public knew, this decision was based upon the results of the reviews which he had commissioned.

The SST Review Committee Report Becomes Public

The report of the interdepartmental SST Review Committee became public at the end of October 1969, as a result of the efforts of Representative Reuss of Wisconsin. He described how he obtained the documents as follows:

I had great difficulty. I first got wind that there was such a report about a month ago, and I thought that the taxpayers of this country had a right to look at it. So I wrote the Administration, "May I, sir, have a copy of this report?" And I got back a letter from the Department of Transportation saying, "This is privileged. You can't see it. You're just a Congressman."

Well, I took this up with our Freedom of Information subcommittee and they pointed out that this squarely violates an agreement that the President made, which is that only the President can claim privilege, not the Department of Transportation, or anybody else. And with that, their house of cards collapsed and I got the report. And now I see why they didn't want to give it to me, because it completely contradicts everything they said and renders this one of the worst fiascos in our sorry history of waste.[7]

Representative Reuss in fact got much more than the review committee's panel reports. In addition, he received copies of a draft summary report, together with letters from members of the review committee to the chairman protesting this summary.[8] These documents suggest not only that Congress and the public were misled about the technical basis of agency decisions, but also that an effort was made to mislead the President about the Committee's conclusions.

Because of the insights this episode provides into the ways in which government "channels" sometimes work, we will discuss it at some length here. Following the completion of the panel reports, the chairman of the Committee, Undersecretary of Transportation James Beggs, wrote a summary report and circulated it to the other members of the committee, requesting their comments within twenty-four hours. Both the biased nature of Beggs's summary and the haste he required for responses provoked a storm of protest from Committee members.

The treatment of the concerns raised by the panel reporting on the environmental impact of the SST may give an indication of the reasons for their consternation. Beggs summarized their conclusions as follows:

[The SST] has the potential for further deteriorating the environment in the environs of the airport and within the area encompassed by the sonic boom path (on the ground) when the aircraft is flown supersonically. However this potential was not considered to be a deterrent to the SST program; instead, when and if it did move forward, this potential should be considered in detail and resolved as early as possible.

. . . Increased water vapor released into the atmosphere from combustion of aircraft fuel could be a problem in terms of local climate and changes in atmospheric circulation and must be further examined.

The foregoing environmental factors are potentially serious and therefore should not be overlooked and underestimated. They are largely known, and can be carefully examined, and a decision made to avoid them.[9]

Contrary to this statement, most of the panel *had* found the consequences for the environment to be a "deterrent to the SST program." Moreover, Beggs's statement in his draft summary that a decision could be made to avoid the environmental problems flew in the face of the environmental panel's report. For example, as Lee DuBridge, science advisor to the President, pointed out to Beggs, there was no practical way to avoid the sonic-boom problem. He also stated that he was doubtful that engines could be designed which were sufficiently light and powerful to be adequate for the SST and also sufficiently quiet to avoid the airport noise problem.

Hendrik Houthakker, member of the Council of Economic Advisors and chairman of the ad hoc SST Review Committee's Economics Panel, expressed what appeared to be a virtually unanimous criticism of Beggs's summary:

It does not adequately reflect the views of the working panels and the members of the Committee. It contains primarily the most favorable material, interspersed with editorial comments, and thus distorts the implications and tenor of the reports.[10]

It appears from reading the letters of protest that Beggs was also violating an explicit commitment which he had made that the committee as a whole would present its views to Secretary of Transportation John Volpe. Several of the members of the committee referred to a letter in which Beggs stated that

after these working panel reports have been received, reviewed, and accepted by the Committee, we will collectively make our views known to Secretary Volpe, who in turn will make his recommendation to the President."[11]

All we know concerning the results of the protests is that a meeting between the full committee and Secretary Volpe was arranged a few days later. We do not know how the committee's views were ultimately presented to the President. This question apparently also bothered the chairman of the committee's environmental panel, Assistant Surgeon General Charles C. Johnson, Jr., who requested in a letter to Beggs that

the collective recommendations to be submitted to President Nixon . . . be provided to the members of the committee and the panels. This would afford the participants an opportunity to learn how their views have been interpreted and whether their efforts have indeed been useful.[12]

The Release of the Garwin Report

Garwin was asked to testify before the House Appropriations Committee in April 1970—a year after the completion of the Garwin Report.[13] This request was followed by invitations to testify before a number of other Congressional Committees.

Garwin's testimony was quite damaging to the administration's case for the SST—particularly his revelations of the magnitude of the airport noise problem and the extent to which the design that the Nixon administration had accepted fell short of the original contract specifications. It should be understood, however, that Garwin continued to respect the rules of confidentiality of the executive branch. He refused to tell Congress anything about his panel's report or even the membership of the panel. He was only willing to give what he carefully identified as his own personal opinions, documented by reference to public documents. In an interview, Garwin explained his view of the advisor's responsibilities as follows:

I'm not a full-time member of the administration and I feel like a lawyer who has many clients. The fact that he deals with one doesn't prevent him from dealing with another so long as he doesn't use the information he obtains from the first in dealing with the second. Since there are so few people familiar with these programs, it is important for me to give to Congress, as well as the administration, the benefit of my experience.[14]

Meanwhile Representative Reuss had asked the President's science advisor to release the Garwin Report—citing once again the Freedom of Information Act, as he had in the case of his request for the SST Review Committee report. This time his request was refused, however. The situation was somewhat different in that the Garwin Report had been commissioned by the Executive Office of the President, while the SST Review Committee's report had been officially commissioned by the Secretary of Transportation. As a consequence, the Nixon administration apparently felt that a stronger argument could be made that the Garwin Report fell under the protection of executive privilege.

After Representative Reuss asked for the Garwin Report and had been refused a second time, a suit was filed calling for its release under the Freedom of Information Act.[15] On a governmental motion, the suit was dismissed in District Court on the grounds that the Garwin Report was indeed protected by executive privilege. This decision was unanimously reversed on appeal, however, and the case was remanded to District Court for trial on its merits. These preliminary skirmishes had consumed more than a year, however, and events had outrun the slow judicial process. By the summer of 1971, the fate of the SST program had been decided by Congress without the benefit of access to the Garwin Report. Thus, further suppression of the report could serve the administration no very important purpose. Loss of the case by the government, on the other hand, would set a precedent adverse to the administration—by putting teeth into the Freedom of Information Act. Thus, on August 17, 1971, Edward David, Jr., the new Presidential science advisor, released the report. In his covering letter he blandly told the plantiffs:

Our compliance with your request will moot any further litigation.... Our action in this regard has been prompted by continued public interest and certain impressions which have arisen depicting the government as attempting to conceal hitherto undisclosed factual data on the SST program. To dispel any further misconceptions that might result from continued litigation, we are releasing the report at this time.

In connection with its release, I would like to place the report in proper perspective so that there can be no misunderstanding about its role in the formulation of the Administration's position on the SST program. The report was one part of a full consideration of the program in early 1969. Other reviews recommended continuation of the program in contrast to one recommendation of this report.[16]

When your authors wrote Dr. David requesting a list and/or copies of the positive reports which he mentioned in this letter, we received no reply.

Thus ends our tale of how the Nixon administration tried to keep from the public the unfavorable results of its comprehensive reviews of the SST program. We now turn to a consideration of two reports relating to the SST program which were voluntarily released.

The NAS-NRC Report on Sonic Boom Effects

The National Academy of Sciences' National Research Council (NAS-NRC) received in 1964 a contract by the Federal Aviation Administration to set up a committee to monitor the federal government's sonic-boom research program. In 1968 the NAS-NRC Committee on SST-Sonic Boom issued a series of reports on the subject. One of these reports—that dealing with the effects of sonic booms on buildings—is the focus of our concern here.

The conclusion of this report stated that "the probability of material damage being caused by sonic booms generated by aircraft operating supersonically in a safe, normal manner is very small."[17] The *New York Times* headlined the resulting story: "Sonic Boom Damage Called 'Very Small'; Wider Study Urged."[18]

In fact, the committee's conclusion should be read to mean that the probability of a single boom damaging any particular building was small—not that the total damage would be slight. It is clear that a fleet of several hundred SSTs flying continuously over the United States would cause a trillion (10^{12}) such individual events per year. Simple calculations based on extensive government test results lead to the estimate that, although damages would average only a fraction of a cent per event, total damages would be on the order of a billion dollars each year.[19] This was obviously the point of interest to the public, yet the NAS-NRC Committee did not make it, and the public was misled by articles such as the *Times* story referred to above.

This case has a particularly interesting sequel because a serious effort was made by an independent scientist to set the record straight.[20] William Shurcliff, the physicist who founded and directed the Citizens League Against the Sonic Boom, had made public estimates of the considerable sonic-boom damage from a fleet of SSTs flying supersonically over land. He consequently became quite concerned that the conclusion of the NAS-NRC report, carrying with it the prestige of the National Academy of Sciences, would be seen as discrediting the SST opponents. Shurcliff therefore joined with John Edsall, a member of the Academy, in requesting from NAS-NRC a public statement clarifying or correcting the report. This proved surprisingly difficult to obtain.

The two scientists began by writing and then telephoning the chairman of the NAS-NRC SST-Sonic Boom Committee, John Dunning, then Dean of the Columbia University School of Engineering. When neither these efforts nor letters to other committee members resulted in any action, Shurcliff and Edsall reluctantly decided to take the matter up with the governing board of the NAS. Finally—still having obtained no public clarification—Edsall circulated a petition among the entire membership of the Academy.

This move finally galvanized the governing board to action: the board issued a circular to the membership conceding that the meaning of the offending sentence (quoted above) could be construed as Shurcliff had construed it while asserting that Shurcliff's was the "only technical criticism" of the report that had been received. They apparently ignored the fact that many major newspapers and even NAS's own *News Report* (March 1968, p. 6) had made the same misinterpretation that had concerned Shurcliff.

Despite the governing board's attempt to mollify the critics, 189 out of approximately 500 NAS members signed Edsall's petitions requesting a public clarification. Other members wrote Shurcliff and Edsall privately, expressing their support. One member, himself a government official, sent the following comment based on his familiarity with the origin of the NAS-NRC SST-Sonic Boom Committee:

I was a member of the Governing Board of the National Research Council of the National Academy of Sciences when the request from the Government for the Academy to make this sonic-boom study was first considered. The whole affair was presented and handled in an atmosphere of secrecy and intrigue. So much was this the case that during the discussion I stated that it did not appear to me that the Academy's advice was being sought on what damage was likely to be produced by the booms from a supersonic transport, or whether such a transport should be built—that decision was apparently already a *fait accompli*— rather, the Academy was being asked to do a "whitewash job" on a publicly unpalatable undertaking.

All information on this subject which has come to me subsequently is consistent with that original judgement. My compliments are accordingly tendered to you gentlemen for courageously taking a position in defense of the public interest with regard to the question which the Academy should have taken, but didn't.

Although this member saw the SST-sonic-boom study as a deliberate "whitewash job," the true explanation for the deceptive way in which the SST-Sonic Boom Committee's report was written may be less blatant—and more insidious. The problem may have originated in the cordial relations which usually exist between advisory committee staff and members and the agency whom they advise. These relations sometimes become so close that we may find the committee's staff ghost-writing the agency's requests for studies,[21] on the one hand, and on the other hand agency officials participating informally in the selection of advisory committee members[22] and in the final drafting of committee reports. It should come as no surprise that in such a system a premium is put on making reports inoffensive to the contracting agency and that the reports consequently are sometimes totally misleading.

In the end, the NAS *News Report* printed its own weak "clarification" of its 1968 news story on the SST-Sonic Boom Committee report, stating in part that

experience has . . . shown that some property damage can be anticipated when such planes fly over populated areas.[23]

No truly public clarification was ever issued. Nevertheless, considerable good may in the end have resulted from the vigorous efforts of Shurcliff and Edsall, for the fuss over the SST-Sonic Boom Committee's misleading report was a major inspiration to the NAS leadership in establishing a new and much more substantial review procedure for NAS-NRC reports.

The SST Community Noise Advisory Committee Report

In the summer of 1970, after Congressional testimony by Richard Garwin, Laurence Moss, and others had made clear the problem of the tremendous noise that SST engines would make at takeoff, the Department of Transportation set up an SST Community Noise Advisory Committee to consider the problem.

Meanwhile, the anti-SST forces continued to use the airport-noise problem effectively as an argument against the SST program. In the early spring of 1971, as the final Congressional votes became imminent and it appeared that the development program might be canceled, Boeing suddenly announced that a number of modifications had been made in the SST design which reduced the airport noise to acceptable levels. Simultaneously, William Magruder, Director of the SST Development Program within the Department of Transportation, released to the media a statement from the chairman of the SST Community Noise Advisory Committee:

We conclude that the level of technology demonstrated by Boeing and General Electric [the contractor for the SST engine] is sufficient to achieve the noise level objectives we recommended [i.e., the same as for four-engine, intercontinental subsonic transport aircraft].[24]

The advisory committee was not asked and did not report what the impact of these changes would be on the SST economics. An indication of the magnitude of this impact became available when Christopher Lydon of the *New York Times* learned that the principal design change was the addition of noise suppressors weighing about 50,000 pounds[25]—a weight nearly equal to the entire payload of the previous design. Another indication of the economic nonviability of the new design came a few months later when, after further SST appropriations had been voted down by Congress, Boeing gave its terms for restarting the development program. Included in its new cost estimate was $350 million for the development of an entirely new and quieter SST engine.[26]

With this example we conclude our presentation of case histories of how administration officials carried out their responsibilities in passing on to concerned Congressmen and citizens the technical analyses of the SST program which they had received. This sorry record is an important piece of evidence we offer to support our argument that Congress and the public need their own science advisors.

NOTES

1. The panel reports, Undersecretary of Transportation James Beggs's summary, and the letters by members of the committee protesting the contents of this summary were entered into the *Congressional Record* 115 (1969): 32599-32613 by Representative Yates of Illinois.

2. "Final Report of the Ad Hoc Supersonic Transport Review Committee of the Office of Science and Technology," March 30, 1969, reprinted in the *Congressional Record* 117 (1971): 32126-32129.

3. Ibid., p. 32126.

4. Ibid., p. 32126.

5. Ibid., p. 32128.

6. Ibid., p. 32129. Contrast these statements in the Garwin Report with the following comment by Presidential science advisor Lee DuBridge, in a letter to Rep. Henry Reuss

dated April 3, 1970: "It would be unfortunate to leave the impression that the [Garwin report] was 'highly critical' of the SST program." [Reprinted in *Congressional Record* 117 (1971): 32125.] DuBridge's letter stated further that the Garwin Report was prepared at President Nixon's request and would not be released; the quoted statement was evidently intended to deceive Reuss as to the report's actual conclusions.

7. NBC radio interview with Rep. Reuss, reprinted in *Congressional Record* 115 (1969): 34743.

8. *Congressional Record* 115 (1969): 32599-32613.

9. Ibid., p. 32606.

10. Ibid., p. 32608.

11. Ibid., p. 32610.

12. Ibid., p. 32607.

13. U.S. Congress, House, Committee on Appropriations, *Department of Transportation and Related Agencies Appropriations for 1971, Part 3,* April 23, 1970, pp. 980-994.

14. Quoted in *Saturday Review,* August 15, 1970.

15. The suit was filed by the American Civil Liberties Union on behalf of Gary A. Soucie, executive director of the Friends of the Earth, and W. Lloyd Tupling, Washington representative of the Sierra Club. Peter L. Koff of Boston was the volunteer attorney.

16. Letter from Edward E. David, Jr., President Nixon's science advisor, to Peter L. Koff, August 17, 1971.

17. Subcommittee on Physical Effects, NAS-NRC Committee on SST-Sonic Boom, *Report on Physical Effects of the Sonic Boom* (Washington, D.C.: National Academy of Sciences, February 1968).

18. *New York Times,* March 5, 1968. (This story was run in early editions but removed from the final edition of this date for space reasons. We received a copy of the article from the *New York Times* morgue.)

19. Results of government tests over a number of cities with military jets compiled by William Shurcliff in his *SST/Sonic Boom Handbook* (New York: Ballantine, 1970) give an average of about $600 damage awards per million "man-booms"—even for sonic booms considerably less intense than those which would accompany the proposed SST. If we then assume that each of 400 SSTs flies 10,000 miles a day at supersonic speeds, creating a 50-mile-wide boom path populated with the average density in the forty-eight contiguous states of about 60 people per square mile, we obtain a rough estimate of 5 trillion man-booms per year and $3 billion annual damage. This calculation is obviously very approximate. We should also note that the National Bureau of Standards used the same damage awards figures compiled by Shurcliff, but used the population of the relevant metropolitan areas—instead of cities—over which the boom tests were conducted, and thus obtained an estimate of $222 damage per million man-booms. [U.S., Environmental Protection Agency, Report no. NTID 300.12, *The Effects of Sonic Booms and Similar Implosive Noise on Structures* (Washington, D.C.: Environmental Protection Agency, December 31, 1971).] The conclusion remains unassailable that if SSTs were flown over the United States, the damage to structures from sonic booms would be very costly. And the legal costs could dwarf the actual damage costs—see W. F. Baxter, "The SST: from Watts to Harlem in Two Hours," *Stanford Law Review,* November 1968, pp. 1-57.

20. The following discussion is based on information in Dr. Shurcliff's files.

21. John Walsh, *Science* 172 (1971): 242.

22. An example is discussed by Nicholas Wade, *Science* 173 (1971): 610.

23. *NAS-NRC-NAE News Report,* February 1969, p. 11.

24. Reported, for example, by Christopher Lydon in the *New York Times,* March 1, 1971, p. 15. We obtained a copy of the report itself from Dr. Leo L. Beranek, chairman of the SST Community Noise Advisory Committee. It was dated February 5, 1971, and was in the form of a one-page memorandum addressed to William Magruder.

25. *New York Times,* March 1, 1971, p. 15.

26. *Washington Post,* May 20, 1971, Sec. A, pp. 1, 8.

CHAPTER 5

Invoking the Experts: The Antiballistic Missile Debate

> *. . . the report sent to the Secretary of Defense said that this equipment will do the job that the Department of Defense wants to do. . . .*
>
> —John Foster, Director of Defense Research and Engineering, citing secret O'Neill committee report on the Safeguard ABM system.
>
> *Dr. Foster's remarks indicate that we made recommendations that in fact we did not make.*
>
> —Professor Sidney Drell, member of the O'Neill committee.

In the previous chapter we presented some examples of the ways in which the public can be misled by the selective release and suppression of analyses and information on which government decisions are based. In this chapter we consider a debate during which government officials publicly misrepresented confidential advice. The advice concerned the effectiveness of first the Sentinel and later the Safeguard antiballistic missile systems.

Background

The search for a defense against intercontinental ballistic missiles armed with nuclear explosives began even before the development of the offensive weapons had been completed. The first contracts for feasibility studies

59

on an antiballistic missile (ABM) system were let by both the U.S. Army and Air Force in 1955.[1]

Two years later, in October 1957, the launching of the first artificial earth satellite (*Sputnik*) by the Soviet Union convinced most Americans with a dramatic suddenness that the Soviets had developed a capability for intercontinental nuclear missile warfare.

The Armed Services responded to the resulting tremendous concern by proposing the deployment of an ABM system. On November 20, 1957, less than two months after the launching of *Sputnik,* the *New York Times* reported that Army Chief of Staff Maxwell Taylor made a proposal to the Joint Chiefs of Staff that the Army antiaircraft missile system be upgraded into a system with ABM capabilities over a period of three years and at a cost of $6.7 billion. The next day the *New York Times* reported that the Air Force had submitted a position paper to the Joint Chiefs which threw doubt on the capabilities of Army's proposed system. A few days later the Air Force announced that it was developing its own ABM system.[2]

PSAC is Created

The decision in this case was not entirely up to the military, however. In response to the crisis triggered by the launching of *Sputnik,* President Eisenhower had turned for advice to scientists and engineers outside the government. Most of these outside experts had become involved with weapons technology during World War II, when they had gained the nation's respect by leading the efforts which resulted in the development of radar and nuclear weapons. After the war they had remained advisors to the Atomic Energy Commission (AEC) and the Department of Defense. A month after *Sputnik,* Eisenhower gave them direct access to the White House by moving the Science Advisory Committee of the Office of Defense Mobilization into the White House as the President's Science Advisory Committee (PSAC). The president of MIT, James Killian, served as PSAC's first chairman and also as the President's full-time science advisor.

After consulting with PSAC, President Eisenhower decided not to approve the deployment of an ABM system—on the grounds that the technology was inadequate. Instead, following PSAC's advice, he created the new civilian post of Director of Defense Research and Engineering to supervise the armed forces' research and development activities. The first person appointed to the new post was a member of PSAC, Herbert York, a physicist and the director of the AEC's nuclear weapons development laboratory at Livermore, California.

But the Democratic majority in Congress blamed the Soviet space triumph on the complacency of the Eisenhower administration and was not satisfied with these actions. Congressional committees were set up to investigate the situation.

The chairman of the Senate committee, Lyndon B. Johnson—then a Democratic senator from Texas and the Senate's majority leader—was particularly critical of the decision not to develop an ABM system.[3] The United States succeeded in launching its own satellite a few months after the Soviets, however, and the criticism eventually subsided.

In 1960, as the Presidential election approached, the issue came alive again. And in October, just before the election, the Democratic Presidential candidate, Senator John Kennedy, in a speech to an American Legion audience, denounced the Eisenhower administration for having allowed a "missile gap" to develop and for its failure to deploy an ABM system.[4] After Kennedy was elected, however, his science advisors quickly convinced him that the technology was still inadequate, and he refused to order deployment despite a continuing public debate, fueled in part by Soviet claims of breakthroughs in their own ABM development program[5] and in part by opponents of the proposed nuclear test ban who seized upon the danger of the Soviets winning the "antimissile missile race" as a reason for continued atmospheric testing.[6]

Occasional public statements during this period indicated a parallel debate going on within the executive branch between the scientific advisors and the generals. In January 1962, Hans Bethe, one of the most eminent scientific advisors on strategic weapons, stated that he felt that development of an effective antimissile missile was hopeless.[7] A few months later General Barksdale Hamlett, Vice Chief of Staff of the Army, argued the opposite view.[8] In March 1963, General Maxwell Taylor, now Chairman of the Joint Chiefs of Staff, warned in Congressional testimony that the United States must win the race for an antimissile missile.[9] At the same time, however, the Department of Defense undertook a major program to develop multiple warheads for U.S. strategic missiles in order to insure that the United States would be able to overwhelm any Soviet ABM system by sheer force of numbers. The scientific advisors argued that the Soviets could similarly penetrate any U.S. missile defense with multiple warheads or other "penetration aids."[10]

The year 1964 was again a Presidential election year, and the Republican candidate, Senator Barry Goldwater, launched an all-out attack on the reliability of the U.S. missile deterrent and the lack of progress of the ABM development program. He was engagingly candid in stating that he was encouraged to make this attack by the fact that John Kennedy had used the "missile gap" charge with considerable effect against the Eisenhower Administration.[11] Goldwater's attack had little impact, however, as the major issue of the campaign became the war in Vietnam.

In late 1965 the Joint Chiefs of Staff, apparently discouraged with the political prospects of an ABM system oriented toward the Soviet Union, recommended deployment of an anti-Chinese system. (The Chinese had tested their first nuclear device a year before.[12]) But President Johnson, apparently strongly influenced by the impact which the $20 billion program would have had on a budget already strained by the Vietnam War and "Great Society" programs, sided with Secretary of Defense Robert McNamara against deployment.

The Pressures for Deployment

The pressure for deployment continued to mount. In November 1966, Secretary of Defense McNamara made public the information that the Soviet Union was deploying an ABM system. According to Defense Department leaks, after the initial deployment of one ABM system around Moscow, deployment of another system had begun across the routes which U.S. missiles would travel in an attack on the Soviet Union. The Senate had already in the spring of 1966 added $167.9 million to the Defense budget to be used for ABM "preproduction funds." The funds had not been requested by the administration, and they were not spent. Secretary McNamara responded to the heightened pressures for deployment by revealing more about the multiple warheads which were being developed for U.S. missiles to guarantee penetration of any Soviet system. Later it became clear that the larger Soviet "ABM system" was actually an antiaircraft system.[13]

In 1967, as his political position became weaker, President Johnson's support for McNamara's anti-ABM position also weakened. In his annual budget message to Congress, Johnson asked for funds for the deployment of a U.S. ABM system in case an agreement with the Soviets for a mutual moratorium on deployment could not be achieved.[14]

This weakening of the President's stance triggered an all-out public campaign for the ABM by the Joint Chiefs. Their chairman, General Earl Wheeler, stepped so far out of his role as McNamara's subordinate that he presented the case for ABM deployment on television.[15]

At about this time McNamara made a last attempt to convince President Johnson of the folly of going ahead with the deployment of an ABM system. He invited all the men who had served as Presidential science advisors or as Directors of Defense Research and Engineering (DDRE) to meet with Johnson and to present to him their views on the proposal for deployment of an American ABM system. All except the incumbent Director of Defense Research and Engineering, John Foster, told the President their reasons for opposing such a move. Johnson was not impressed.[16]

The Decision to Deploy

The pressure on the administration increased further that autumn when key Congressional committees joined the Joint Chiefs in calling for a decision to deploy ABM. The Senate Appropriations Committee under Senator Richard Russell (D.-Ga.) publicly informed the President that his administration would have to bear the responsibility for any further delay.[17] And Senator John Pastore (D.-R.I.), chairman of the Joint Committee on Atomic Energy, announced that his committee would also fight for deployment.[18]

The coup de grace was delivered by Republican Presidential aspirant Richard M. Nixon on September 14, 1967. He stated that, unless Johnson decided to deploy the ABM, the President would find the issue of the "missile gap" turned upon him during the forthcoming 1968 Presidential campaign. "It's a deadly boomerang," he gloated.[19]

This time Johnson was on the wrong side of the Vietnam issue and in no position to take such a threat lightly. On September 18, four days after Nixon made his statement, Secretary of Defense McNamara announced the administration's decision to deploy a "light" anti-Chinese ABM system. The speech in which he made this announcement ironically also presented an extremely effective argument against deployment and warned against further surrender to the pressures for escalation of the arms race.

There is a kind of mad momentum intrinsic in the development of all nuclear weaponry. . . . The danger in deploying this relatively light and reliable Chinese-oriented A.B.M. system is going to be that pressures will develop to expand it into a heavy Soviet-oriented system.[20]

McNamara's announcement marked the end of an era in the relationship between scientists and the executive branch. Scientists had gained influence—in some cases greater than that of the Joint Chiefs—as a result of the *Sputnik* crisis. A decade later, however, when it was obvious that the United States was far ahead of the Soviet Union in strategic weapons and in space technology generally, this area ceased being one of overriding public concern. The decision-making power then returned to the arms lobby.

Citing the Experts

Just as McNamara's September 18 speech served to mark the end of a decade of unparalleled influence for scientists in United States strategic weapons policy, it also gave an indication of what the new relationship between scientists and the administration in this area was to be. Toward the end of his exposition on the futility of building a heavy ABM system as protection against Soviet strategic missiles, McNamara invoked the names of the scientists whom he had brought together in President Johnson's office:

If we . . . opt for a heavy ABM deployment—at whatever price—we can be certain that the Soviets will react to offset the advantages we would hope to gain.

It is precisely because of this certainty of a corresponding Soviet reaction that the four prominent scientists—men who have served with distinction as the science advisors to Presidents Eisenhower, Kennedy, and Johnson, and the three outstanding men who have served as directors of research and engineering to the three Secretaries of Defense—have unanimously recommended against the development of an ABM system designed to protect our population against a Soviet attack.

These man are Doctors Killian, Kistiakowsky, Wiesner, Hornig, York, Brown, and Foster.[21]

McNamara's statement was misleading in that he presented only half the truth. He failed to mention that all of these scientists (with the exception of Foster) had also opposed the deployment of the Chinese-oriented system which he was announcing. He thus obscured the basic fact that a political and not a technical decision has been made. As skeptics suggested, the primary mission of the ABM system was not to defend against Chinese or even Soviet attacks; fundamentally, it was a Republican-oriented system.

Until McNamara made his announcement, the battle over whether or not to deploy an ABM system was, as we have seen, primarily a battle for the President's mind. Once McNamara and the President's Science Advisory Committee had lost that battle, however, a few of the scientific advisors, notably Bethe, Wiesner, and York, helped take the issue to Congress and the public. We will discuss the public debate which ensued in a later chapter. Here we will only describe some incidents which provided glimpses of the attention accorded within the executive branch to those advisors—notably those then on PSAC—who continued to express their opposition to the ABM within the administration on a confidential basis.

The Senate Foreign Relations Committee Hearings

Much of the technical basis for Congressional criticisms of administration ABM proposals developed during hearings held by a special Subcommittee on International Organization and Disarmament Affairs of the Senate Foreign Relations Committee. Senator J. W. Fulbright (D.-Ark.), chairman of the full committee, set up the subcommittee after the 1968 hearings of the Senate Armed Services Committee—which, following its usual practice (since changed), had not heard a single witness opposed to the administration proposals. The special subcommittee, chaired by Senator Albert Gore (R.-Tenn.), held hearings on the administration's ABM proposals during 1969 and 1970.

The subcommittee conducted its first hearings in March 1969, before the new Nixon administration had taken a public position on the ABM. During these hearings a number of former top scientific advisors on strategic weapons matters, including Bethe, Killian, Kistiakowsky, and York, testified against the Johnson administration's ABM proposal.

The objections of these scientists were of two basic types: technical—they felt that the proposed missile defense could be easily penetrated even by Chinese missiles; and strategic—they felt that the deployment of an ABM system was unnecessary and could trigger a new arms race with the Soviets. As time went on, however, the debate focused more and more on the technical objections. It was obviously the hope of many ABM opponents that the technical arguments would be more effective than arms-race considerations in convincing Congressmen of all political persuasions to oppose the deployment decision.

Secretary Laird's List

When President Nixon, on March 14, 1969, finally announced his decision to deploy an ABM system, it turned out to be basically the Johnson administration's system with a different name, "Safeguard," and with the missile sites moved away from the cities—an obvious response to the opposition which had developed in many suburban areas against having nuclear weapons in their "back yards." (See Chapter 13.)

Nixon's Secretary of Defense, Melvin Laird, came to present this proposal to Senator Gore's subcommittee. But as he was giving his opinion that the deployment of the proposed ABM system would not trigger a new arms race with the Soviet Union, Fulbright interrupted:

SENATOR FULBRIGHT: Of course, I do not think that [the Soviets] are really very bothered about the ABM either, because I am sure they know, as nearly every witness outside the Pentagon knows, it is not much good. We have had a number of scientific witnesses who have said—

SECRETARY LAIRD: I hope you will listen to other scientific witnesses too.

SENATOR FULBRIGHT: I know the Pentagon.

SECRETARY LAIRD: Not from the Pentagon but outside the Pentagon.

SENATOR FULBRIGHT: Are there any outside scientists that are not either in the contracting business, working for your contractors or in your employ? What independent scientists are there? I would like you to name them.

SECRETARY LAIRD: I will be glad to supply you with a list.[22]

When the list came back, it named eight scientists. Senator Gore invited four of them to testify: Detlev Bronk, who had served simultaneously as chairman of the DOD's top science advisory committee, the Defense Science Board, and as president of the National Academy of Sciences; Edward Teller, popularly known as the "father of the H-bomb," who was associate director of the Lawrence Radiation Laboratory at Livermore, one of the AEC's weapons laboratories; Eugene Wigner, Professor of Physics at Princeton, winner of the Nobel Prize for Physics in 1963, and former member of the General Advisory Committee of the AEC; and Gordon MacDonald, former vice-president of the Institute for Defense Analyses, a Defense Department "think tank." (At the time of the hearings MacDonald was Vice-Chancellor for Research and Development at the University of California at Santa Barbara, and a member of both the Defense Science Board and PSAC. He was shortly to be appointed by President Nixon to the new Council on Environmental Quality.)

Of the four scientists, three were willing to testify; Bronk asked in a letter to be excused from testifying, giving as his reason: "my opinions would be dangerously unqualified."[23]

The other three testified but did not attempt to rebut the technical objections of the ABM opponents. Instead they supported the President's decision to deploy the Safeguard ABM system because they saw it as a long-awaited commitment of the nation to the idea of missile defense: Teller and Wigner in particular saw Safeguard as a step toward the development of a "heavy" system which would be designed to defend the U.S. population against Soviet attack.[24] Apparently it did not bother them that President Nixon had specifically rejected the mission of a Soviet-oriented population defense in his deployment announcement, stating his belief that an effort in that direction would only trigger an arms race between Soviet offensive and U.S. defensive forces which the United States could not win.[25]

MacDonald was willing to endorse a very limited deployment of the Safeguard system if it were accompanied by a commitment to develop a system which could actually carry out one of the missions which President Nixon had given the Safeguard system—defense of some of the U.S. Minuteman missile bases against a possible Soviet first strike. MacDonald stated that "if properly emphasized, research and development could, in a short time, produce a system much better suited to defending our strike forces."[26]

At the end of MacDonald's presentation Senator Gore commented:

There is a great similarity between the conclusion at which you arrive and that of Dr. Hornig which he has presented. Your logic is powerful. Thank you very much.[27]

Hornig, formerly President Johnson's science advisor, had just testified against deployment.

It appears that the administration made an exception to its rules of confidentiality in volunteering MacDonald's services as a witness for the Safeguard ABM deployment. The other members of PSAC, who were almost unanimously of the view that the deployment of the Safeguard ABM system was senseless, were requested to keep these views confidential.

Deputy Secretary Packard's Consultations

Following Defense Secretary Laird's testimony before Senator Gore's subcommittee, a more detailed discussion of how the Safeguard ABM system would work was presented by Deputy Secretary of Defense David Packard. Packard had had the responsibility of directing the two-month-long review within the Nixon administration which resulted in the modified Safeguard ABM deployment proposal.

Toward the end of Packard's testimony, Fulbright asked for more information about who had participated in the review:

SENATOR FULBRIGHT: I think it would be very interesting to have before the subcommittee just who participated in the review and how, and in what depth it was made. The reason that particularly appeals to me is that this committee has done some reviewing too, with some of the leading authorities in the field of nuclear warfare. . . .

MR. PACKARD: The review utilized the full staff of the Defense Department, and those people that the Department had utilized for scientific evaluation. In addition to that, I have talked to some scientific people on my own about the matter, some people who have no connection with the—

SENATOR FULBRIGHT: Who were they who had no connection with the Pentagon? There is nothing classified or secret about this sort of thing is there?

MR. PACKARD: One of the men that I talked to, I have a very high regard for, is Professor Panofsky.[28]

When Senator Fulbright asked the names of the other outside scientists Packard had consulted, he couldn't remember but promised to send Fulbright a list.

Two days later Panofsky appeared in response to an invitation to testify. A physicist and Director of the Stanford Linear Accelerator Center, Panofsky had been some years before the chairman of PSAC's Strategic Weapons Panel and was still involved in advising the executive branch on these matters. He had not (to the authors' knowledge) previously made public his views on the ABM.

Dr. Panofsky began as follows:

. . . To clarify the record I would like to state that I did not participate in any advisory capacity to any branch of the Government in reviewing the decision to deploy the . . . Safeguard system—I appreciate having had the opportunity of an informal discussion with Mr. David Packard, Deputy Secretary of Defense, several weeks ago prior to the . . . decision.

SENATOR GORE: To what extent was this? Was there an extended conversation over a period of time?

DR. PANOFSKY: About half an hour . . .

SENATOR GORE: Did you call upon him or did he call upon you?

DR. PANOFSKY: We happened to accidentally meet at the airport.[29]

Panofsky thereupon went on to detail at considerable length his reasons for believing that the Safeguard ABM system deployment decision was "an unwise decision from many points of view, from the point of view of sound engineering judgment, economy, and stopping the arms race."[30]

If this was the extent of consultation that Deputy Secretary of Defense Packard had had with Dr. Panofsky and this the type of advice that he had received from him, what about the list of other outside consultants he had promised Senator Fulbright?

When the list arrived it was entitled, "List of Scientists and Engineers Consulted by [Director of Defense Research and Engineering] Foster on ABM."[31]

Dr. Foster's Consultations—1969

The scientists listed as having been consulted by Foster were: (1) the members of the President's Science Advisory Committee; (2) the members of Foster's own advisory committee, the Defense Science Board; and (3) the members of the Defense Science Board Task Force on ABM. But when Fulbright followed up with a list of written questions to Packard asking for more details about the consultations, the replies revealed that (1) the President's Science Advisory Committee was "consulted" three days *after* Nixon had announced the Safeguard deployment decision; (2) the Defense Science Board had not been consulted at all during the review process; and (3) the Defense Science Board Task Force on ABM was consulted only once—at the end of the two month review process,[32] three days before Nixon's announcement of the deployment decision.

Thus it became clear, despite the Defense Department's best efforts, that the outside scientists who had been so influential in helping to shape the Nation's strategic weapons policies for a decade had been almost entirely excluded from the Nixon administration's ABM review process. Indeed, all outside review was excluded. It appears, from the reluctance that was shown in admitting this fact, that the administration was more willing to forego the advice than it was to forego the support for the ABM which could be obtained by invoking the names of prominent advisory committees.

Dr. Foster's Consultations—1970

In 1969 the Senate authorized appropriations for construction of the first two bases of the Safeguard ABM system as a result of Vice President Agnew's tie-breaking vote. A year later the Nixon administration was asking for funds for additional sites. Once again Senator Gore's subcommittee held hearings.

This time three former Presidential science advisors (Kistiakowsky, Wiesner, and Hornig), a former Director of Defense Research and Engineering (York), and Panofsky were among those who presented the technical arguments against expansion of Safeguard.

The technical case for the administration was presented this time by Dr. Foster, Director of Defense Research and Engineering. Foster had not gotten far into his testimony, however, when Senator Fulbright confronted him as he had Secretary Laird and Deputy Secretary Packard the previous year, with the impressive list of experts who had testified against further deployment of the Safeguard system:

> What concerns me is the fact that there are so many scientific authorities in the United States, those not in the employ of the Defense Department, and many people who are not scientists, but who are knowledgeable about Soviet relations and have studied them for many years, and also have studied disarmament matters who think [further deployment] endangers the success of the SALT [Strategic Arms Limitations] talks. . . . You also know that every former Presidential science advisor is opposed to expanding Safeguard at this time.[33]

Fulbright then went on to list some recent Department of Defense fiascos with advanced weapons systems. Some of these systems had cost billions of dollars more than the department had originally told Congress, and the performance of many had fallen so far short of specifications that it was not clear whether they could be used at all. He then continued:

> In view of this record, I don't see how you can be so confident of your judgment about these matters. It really shakes my confidence as to whether the Department is capable of an objective view of these matters.[34]

Foster was stung into making a rebuttal:

DR. FOSTER: Mr. Chairman, you have indicated the number of scientists who oppose this Safeguard deployment.

SENATOR FULBRIGHT: There are several grounds. They oppose it on the SALT talks alone. Then in addition they oppose it on the ground that it isn't technically feasible, at the present time at least.

DR. FOSTER: Well, Mr. Chairman, let me just simply point out that I asked a group of scientists to come together as an ad hoc committee and, before the Secretary of Defense made his recommendation to the President, review the program. I deliberately chose scientists who opposed the deployment of Safeguard as well as those who favored it.

In fact, as I recall, when they met there were more against it than for it. I had, however, one very simple instruction for them—to put politics aside and just ask the question: Will this deployment, with these components, do the job that the Department of Defense is trying to do? . . .

There was considerable concern about this move, but *the report sent to the Secretary of Defense said that this equipment will do the job that the Department of Defense wants to do.* . . . [Emphasis added.]

I think it is extremely important that, when you ask a scientist for his opinion, you make sure that you have found a way to rule out political factors, because, as you and Secretary Laird noted at our last hearing, the scientist doesn't have special competence in that area.[35]

Here Foster appeared to be claiming that the Senators had not been successful in forcing the scientists who had testified before them to keep their political beliefs from biasing their technical presentations. He also indicated his belief that he, an expert himself, had succeeded where the Senators had failed and that, when separated from politics, the technical considerations had turned out to favor the Safeguard system.

When asked to name the members of the ad hoc committee, Foster could not remember all of the names. Among those he mentioned, however, were Drs. Marvin Goldberger and Sidney Drell.[36] These scientists had in turn succeeded Panofsky as chairman of PSAC's Strategic Weapons Panel.

When the Senators asked to see the ad hoc committee report, they were told that it was confidential. Matters did not end here, however, because both Drell and Goldberger wrote to Senator Gore about Foster's representation of the conclusions of the ad hoc committee report (commonly identified as the O'Neill Report after the committee's chairman, Dr. Lawrence O'Neill, president of the Riverside Research Institute, an ABM contractor). Goldberger wrote:

> I can only presume that the implication [was] that our panel supported the arguments presented by Dr. Foster and the Department of Defense in justifying the next phase of Safeguard to your committee.
> *The report took no such position.* [Emphasis in original.][37]

Drell similarly wrote that "Dr. Foster's remarks indicate that we made recommendations that in fact we did not make."[38]

Senator Gore of course invited both men to testify before his subcommittee. A few excerpts will give the flavor of their opinion of the Safeguard ABM system.

DR. GOLDBERGER: ... I assert that the original Safeguard deployment and the proposed expanded deployment is spherically senseless. It makes no sense no matter how you look at it.[39]

... If there are enough highly accurate, large payload Soviet missiles to threaten Minuteman without any defense ... Safeguard is irrelevant.[40]

... The Chinese will be designing their offensive missile force in the face of our emplaced system whose operating characteristics will be precisely known. Since they are not noted for their stupidity, they will in all probability take steps to counter the defense by the use of penetration aids, or circumvent it entirely by, say, attacking Hawaii if they just want to kill people or using aircraft or ships to attack West Coast cities with nuclear weapons.[41]

DR. DRELL: ... [Safeguard] simply fails to respond to the threats postulated by the Pentagon, and furthermore it is not cost effective.[42]

SENATOR [CLIFFORD] CASE [D.-N.J.]: ... Your whole opposition to Safeguard is not in any way based upon any contempt or downgrading of... Soviet capability?

DR. DRELL: No sir. It is merely a contempt for the capability of Safeguard.[43]

This, then, was a sample of the anti-ABM opinion on PSAC which the Nixon administration had chosen to conceal behind a wall of confidentiality in 1969 while offering Congress instead the ambivalent endorsement of Dr. MacDonald.

Release of the O'Neill Committee Report

After the devastating testimony of Drell and Goldberger, the Defense Department had little to gain by keeping the O'Neill report secret. The report was released a month later, on July 24, 1970.[44] It addressed the question of how well the Safeguard system would fulfull the missions that President Nixon had assigned it: (1) defense of the U.S. Minuteman strategic missile bases against a Soviet surprise attack (the mission to which the Nixon administration had given the greatest emphasis); (2) defense of the U.S. population against a nuclear attack launched from China (the mission which had originally been given to the system by Secretary McNamara); and (3) "protection against the possibility of accidental attacks from any source"[45] (a mission so ill-defined that it was hardly even discussed).

As to the first mission, the panel concluded:

The group believes that a more cost effective system for the active terminal defense of Minuteman than Phase IIA of Safeguard can be devised.[46]

Regarding the second mission the panel reported a lack of consensus.[47]

When Senator Fulbright put the O'Neill report into the *Congressional Record*, he commented:

[This] is not a ringing endorsement of the Safeguard system. . . .

We have had, in the past, a missile gap. More recently, we have experienced a credibility gap. We seem now to be combining the two in a missile credibility gap which emerges clearly from the record of the Defense Department in attempting to support claims that it has submitted the Safeguard system to independent outside review. The missile credibility gap was opened last year by Mr. Packard's implication that Dr. Panofsky had supported the Safeguard system. It was widened this year by Dr. Foster's assertion that the O'Neill panel had concluded that Safeguard could meet certain objectives. Two members of the O'Neill panel do not agree and surely they must know what they decided and recommended. One of the members of the O'Neill panel, Dr. Drell, went even further and said:

"All analyses of which I am aware make it clear that, if defense of Minuteman is the principal or sole mission of Safeguard, its further deployment cannot be justified."

For we who must rely on the informed judgements of others, as far as technical matters are concerned, Dr. Drell's statement stands as a severe indictment of the Safeguard system and calls into question the tactics employed by the Defense Department in seeking to make it appear that the scientific community supports the Safeguard system as an effective defense of our deterrent missile force.[48]

The Invisible, Inaudible Authorities

We have seen in this chapter how executive branch spokesmen in an important national debate cited the experts while suppressing their reports. The evidence indicates very clearly that for the public to accept such statements at face value is an invitation to governmental corruption of the truth.

In science, the invocation of authority as a substitute for evidence was discredited in the Renaissance. Yet here we find government officials trying to revive this tactic in an effort to deceive the public. It is distressing to see how little criticism of this dangerous tendency has been offered by the scientific community.

Even if the abuses which we have described had not occurred, it would still be against the public interest to conceal the technical bases of public policy. The ABM debate shows that even the general capabilities of advanced strategic systems can be publicly debated without the disclosure of classified details of hardware or tactics. It is characteristic of scientific research that its practitioners are continually testing even the most well-established theories. No scientific statement is protected from question by the eminence of the researcher who has put it forward. Indeed, scientists often gain fame by finding unsuspected imperfections in the edifices raised by their revered predecessors. The technical information which forms the basis for public policy should certainly not be immune from similar reexamination. Although we have in this chapter considered instances where the federal executive branch appears to have had available technically competent advice—even though it did not want to hear it—there are many other instances in which government agencies have received dangerously inadequate or faulty advice. In these cases, some of which will be presented below, it has only been as a result of members of the larger scientific community "raising a ruckus" that government officials have become aware of the inadequacies in their information.

NOTES

1. Jerome Wiesner and Abram Chayes, eds., "Chronology of U.S. ABM Deployment Decisions, 1955-1969," *ABM: An Evaluation of the Decision to Deploy an Antiballistic Missile System* (New York: New American Library, 1969), p. 227.

2. *New York Times*, November 20, 1957, p. 1; November 21, 1957, p. 1; November 30, 1957, p. 1.

3. Ibid., January 24, 1958, p. 1; January 24, 1958, p. 6; July 26, 1958, p. 5.

4. Ibid., October 19, 1960, p. 39.

5. Ibid., November 19, 1960, p. 5; July 11, 1960, p. 4; July 17, 1960, p. 1; December 18, 1960, p. 1.

6. Ibid., September 6, 1961, p. 3; October 31, 1961, p. 15.

7. Ibid., January 6, 1962, p. 4.

8. Ibid., April 19, 1962, p. 22.

9. Ibid., March 30, 1963, p. 4.

10. Ibid., November 9, 1963, p. 1; December 18, 1963, p. 50.

11. Ibid., January 10, 1964, p. 1; January 11, 1964, p. 1; January 16, 1964, p. 1.

12. Ibid., November 24, 1965, p. 1; December 1, 1965, p. 1.

13. Ibid., April 29, 1966, p. 1; November 11, 1966, pp. 1, 18; December 8, 1966, p. 1; February 23, 1967, p. 10.

14. Ibid., January 25, 1967; pp. 17, 20-21.

15. Ibid., February 27, 1967, p. 7; May 4, 1967, p. 1.

16. Elinor Langer, "After the Pentagon Papers: Talk with Kistiakowsky, Wiesner," *Science* 174 (1971): 923 ff; Anne H. Cahn, *Eggheads and Warheads: Scientists and the ABM* (Ph.D. thesis, MIT, 1971), pp. 37, 242.

17. *New York Times*, August 6, 1967, p. 17.

18. Ibid., September 10, 1967, p. 1.

19. Ibid., September 15, 1967, p. 1.

20. Ibid., September 19, 1967. MacNamara's speech is reprinted on p. 18.

21. Ibid.

22. U.S. Congress, Senate, Subcommittee on International Organization and Disarmament Affairs of the Committee on Foreign Relations, *The Strategic and Foreign Policy Implications of Antiballistic Missile Systems*, 91st Congress, 1st Session, March 21, 1969, pp. 204-205. This reference will hereafter be referred to as *ABM Systems*.

23. Ibid., May 21, 1969, p. 561.

24. Ibid. Teller's statement was on May 14, 1969 and begins on p. 501 and Wigners was on May 21 and beings on p. 551.

25. President Nixon's announcement of the Safeguard ABM decision is reprinted in the *New York Times*, March 15, 1969, p. 16.

26. *ABM Systems*, May 21, 1969, p. 543.

27. Ibid., p. 550.

28. Ibid., March 26, 1969, pp. 306 ff.

29. Ibid., March 28, 1969, pp. 327 ff.

30. Ibid., p. 334.

31. Ibid., March 26, 1969, p. 308.

32. Ibid., letter from Packard dated April 15, 1969, reprinted on pp. 612-613.

33. U.S. Congress, Senate, Subcommittee on Arms Control, International Law and Organization of the Committee on Foreign Relations, *ABM, MIRV and the Nuclear Arms Race*, 91st Congress, 2nd Session, June 4, 1970, pp. 441 ff.

34. Ibid., p. 442.

35. Ibid.

36. Ibid., p. 501.

37. Ibid., letter dated June 11, 1970, p. 522.

38. Ibid., letter dated June 5, 1970, pp. 522-524.

39. Ibid., June 29, 1970, p. 528.

40. Ibid., p. 529

41. Ibid., p. 531.

42. Ibid., p. 554.

43. Ibid., p. 578.

44. The O'Neill Report and correspondence related to its release are reprinted along with Senator Fulbright's comments in *Congressional Record* 116 (1970): 27723-27728.

45. Ibid., p. 27726.

46. Ibid.

47. Ibid., p. 27727.

48. Ibid., p. 27725.

CHAPTER 6

Studies as an Excuse for Inaction: The Saga of 2,4,5-T

Background

In 1962 the publication of Rachel Carson's *Silent Spring*[1] touched off a tremendous debate over the environmental and health impact of the use of pesticides. Among other dangers, she pointed out the likelihood that some of the chemicals being used as pesticides were carcinogenic, teratogenic, and/or mutagenic (capable of producing cancer, birth defects, and/or gene defects, respectively). The subsequent report on pesticides of the President's Science Advisory Committee recommended that tests for these effects be conducted on laboratory animals.[2] Accordingly, in summer 1963 the National Cancer Institute (a division of the federal government's National Institutes of Health) contracted with the independent Bionetics Research Laboratories in Bethesda, Maryland to perform such studies.[3] After the studies had been commissioned, however, the research stretched out over years with no published results.

One of the chemicals which Bionetics was commissioned to study was the herbicide 2,4,5-trichlorophenoxyacetic acid, commonly known as 2,4,5-T. The U.S. Army had tested this chemical during World War II for possible use as a defoliant—i.e., to remove concealing foliage.[4] The war ended before it could be used, however. After the war the chemical was introduced into the domestic market as a weed and brush killer. By 1965 it had become so popular that 13 million pounds of 2,4,5-T were being manufactured annually in the United States.[5]

Army testing of 2,4,5-T as a defoliant continued after World War II, with large-scale field tests being conducted in Puerto Rico and Thailand. Finally, the Vietnam War presented an opportunity for the military use of defoliants. From a small beginning in 1961 their use expanded rapidly until the period 1967-1969, when about 2,500 square miles of South Vietnamese forest were being defoliated yearly—about 90 percent using "Agent Orange," a 50-50 mixture of 2,4,5-T and another popular herbicide, 2,4-D.[6] Because of the density of the jungle and in order to have quick results, about ten times as much herbicide was used per acre in South Vietnam as is recommended for domestic use. Indeed, most of the U.S. production of 2,4,5-T was being dumped on Vietnam, and for a time it was difficult to obtain the chemical for domestic purposes.[7] Production was rapidly expanded, however, and by 1968 about 42 million pounds were being produced annually in the United States—more than double the 1966 figure of 18 million pounds.[8]

The Bionetics Reports

In June 1966, while the use of 2,4,5-T was still increasing in Vietnam, the Bionetics Research Laboratories informed the National Cancer Institute (NCI) that its tests on pregnant mice injected with small amounts of 2,4,5-T resulted in greatly increased numbers of birth defects.[9]

The reaction of the NCI was remarkable. Instead of warning the public or the responsible government agencies of the possible danger, the Institute sent the matter back to Bionetics for further study. Surgeon General Jesse Steinfeld later attempted to justify this action by stating that "at that point we did not know whether the results produced by injection were significant. The 2,4,5-T had not been fed."[10] Bionetics apparently was not pressed for further results, however, and two years passed before a second report was delivered to the NCI. The conclusion: 2,4,5-T was also teratogenic in mice when administered orally.[11]

Still the government hardly stirred. According to Surgeon General Steinfeld's later account, on January 30, 1969,

a special preliminary report on the teratogenicity of 2,4,5-T [was made available] at a meeting of scientists from the National Institutes of Health with representatives of the regulatory agencies, Consumer Protection and Environmental Health Services, the National Academy of Sciences, and the chemical industry, attended also by Drs. Phillippe Shubik and Samuel Epstein [two outside scientists].[12]

The meeting did not result in any action, however. The report was passed on the National Institute of Environmental Health Sciences, which according to Steinfeld then spent nine more months conducting "extensive statistical analyses" on the data.[13] (This assertion mystifies us. Having seen the data, we do not see how it would be possible for a competent statistician to spend more than a few days making all reasonable statistical checks for significance of the Bionetics data.[14]

The Mrak Commission

By 1969 seven years had passed since the publication of *Silent Spring,* and the lack of government efforts to tighten the regulation of pesticide use had become obvious. As a result pressure from environmental groups began to mount, stimulating in turn increased resistance from the chemical industry and the political representatives of agriculture. The debate over the banning of DDT became the principal battleground, and the next development in our story of 2,4,5-T was triggered by an incident in that fight.

In April 1969 the Food and Drug Administration (FDA) seized 34,000 pounds of frozen Lake Michigan coho salmon because the fish contained in their fat higher levels of DDT than the limits set by the FDA for meat. This action angered the Republican governors of the states adjoining Lake Michigan as well as Republican House Minority Leader Gerald Ford (Mich.), in whose district the hapless salmon shipper resided. In response to the protests of these important gentlemen and to the rising level of controversy about pesticides in general, Secretary of Health, Education, and Welfare Robert Finch immediately set up a Commission on Pesticides and Their Relationship to Environmental Health. (The commission became known popularly as the Mrak Commission after its chairman Dr. Emil Mrak, Chancellor Emeritus of the University of California at Davis.)[15] The Mrak Commission set up in turn various panels, one of which, the teratology panel, was concerned with assessing the dangers of birth defects resulting from human exposure to various pesticides.

In August 1969—more than three years after Bionetics Research Laboratories had first reported to the government that 2,4,5-T was teratogenic—the teratology panel of the Mrak Commission asked for Bionetics' findings. The request was refused on the grounds that the analysis was not yet complete.[16] On September 24, the panel was finally given the desired information. According to the cochairman of the panel, Dr. Samuel Epstein, this was accomplished "by pulling teeth."[17] On the basis of Bionetics' findings, the teratology panel of the Mrak Commission later recommended in its report that use of 2,4,5-T and a number of other pesticides which had been shown to be teratogenic "be immediately restricted to prevent risk of human exposure."[18]

The Bionetics Report Becomes Public

It is not clear how long the Bionetics results and the Mrak Commission recommendations would have remained secret had it not been for Anita Johnson, who worked with a group sponsored by consumer advocate Ralph Nader studying the food regulation activities of the FDA during the summer of 1969. In going through FDA files, Miss Johnson happened upon a copy of the preliminary report of the Bionetics findings. In September she mentioned the report to a friend, a graduate student in biology at Harvard, who in turn mentioned it in early October to Harvard biologist Matthew Meselson.[19]

Meselson had been deeply involved in the national debates over the United States' stance on chemical and biological warfare, and was already concerned about the teratogenic potential of herbicides. Furthermore, his attention had been called to disturbing stories in South Vietnamese newspapers claiming extraordinary rashes of birth defects in areas which had been defoliated.[20] But, when he tried to get copies of the Bionetics reports, he was informed that they were "confidential and classified".[21]

Meselson soon got copies of the Bionetics reports via an unofficial route. The implications of their findings seemed so serious to him that he immediately informed Lee DuBridge, the President's science advisor.[22] A few weeks later the Nixon administration somehow learned that reporter Bryce Nelson of the *Los Angeles Times* was about to break the story. On October 29, 1969, Nelson was called by the White House just as he was finishing his article[23] and was told that DuBridge had just released a statement in which he announced that, because of the Bionetics findings, "a coordinated series of actions are being taken by the agencies of Government to restrict the use of the weed-killing chemical, 2,4,5-T. . . . The actions taken will assure safety of the public while further evidence is being sought."[24] The major actions announced by DuBridge were as follows:

The Department of Agriculture will cancel registrations of 2,4,5-T for use on food crops effective January 1, 1970, unless by that time the Food and Drug Administration has found a basis for establishing a safe legal tolerance in and on foods. . . .

The Departments of Agriculture and Interior will stop use in their own programs of 2,4,5-T in populated areas or where residues from use could otherwise reach man.

The Department of Defense will restrict the use of 2,4,5-T to areas remote from the population.[25]

On December 5, the Mrak commission report was released.[26]

Dow Chemical Counterattacks

The Department of the Interior carried out the commitment made for it by DuBridge, terminating the use of 2,4,5-T under its control.[27] By January 1, 1970, however, neither the Department of Agriculture nor the Department of Defense had acted to restrict the use of 2,4,5-T in the United States or in South Vietnam. In response to inquiries both departments justified their inaction by stating that it now appeared probable that a contaminant—2,3,7,8-tetra-chlorodibenzoparadioxin, commonly known as "dioxin"—and not the chemical 2,4,5-T itself, had caused the teratogenic effects observed in the Bionetics tests.[28] Therefore, the argument went, if the manufacturers changed their production techniques to minimize this impurity, continued use of 2,4,5-T would be acceptable. This thesis with which the Departments of Agriculture and Defense justified their inaction had been put forward by the Dow Chemical Company, one of the major manufacturers of 2,4,5-T.

The Dow counteroffensive was organized by Dr. Julius E. Johnson, Dow Vice President and Director of Research and a member of the Mrak Commission. (Such conflict-of-interest situations are not uncommon on government advisory committees.) On November 7 he had presented the dioxin theory to the Commission, but was unable to influence its conclusion that 2,4,5-T is a teratogen. Johnson then met on November 25 with officials of the National Cancer Institute and made arrangements for Dow to conduct a new study of the teratogenicity of 2,4,5-T for the NCI with a sample containing much less dioxin than that used by Bionetics. On December 1 he met with DuBridge and informed him of this agreement.[29]

On January 12, 1970, six weeks after designing the study, Dow communicated its findings to the Department of Health, Education, and Welfare (HEW) and the Department of Agriculture, claiming confirmation of its contention that "purified" 2,4,5-T does not cause birth defects. This claim stimulated scientists at both the Food and Drug Administration and the National Institutes of Health to undertake their own tests of the Dow theory.[30]

On February 24 the results of the government studies were presented in a meeting at the Food and Drug Administration.[31] Contrary to the Dow results, the government studies showed that even purified 2,4,5-T was as potent a teratogen as thalidomide, a sedative whose use by pregnant women in Europe in the period 1954-1962 resulted in the birth of thousands of children lacking complete arms and legs. (The dioxin impurity was found to be up to 100,000 times more potent, however. Since the Bionetics sample contained about 30 parts per million dioxin, the effects of the dioxin and those of the 2,4,5-T which it contained were probably roughly comparable.) The discrepancy between Dow's and the government's tests was subsequently partially explained by the facts that: (1) the Dow experimenters administered dosages of 2,4,5-T considerably smaller than those used in the government tests and in most of the Bionetics tests, and (2) Dow scientists had redefined for their own purposes the meaning of the term *teratogenic* to exclude certain effects which the government scientists considered to be birth defects.[32]

It should be noted that it took the government and Dow scientists only six weeks each to execute experiments designed to test the theory which Dow had put forward in defense of continued use of 2,4,5-T. These tests were essentially identical to the Bionetics study, the completion of which had been delayed more than three years by the sponsoring governmental agency after the preliminary results had given evidence of a potentially serious public health hazard. It is hard to imagine better evidence that the government had dragged its feet on the Bionetics results than the almost unseemly haste with which it moved when the possibility was raised that the suspected chemical might be exonerated.

The Congressional Investigation

Both the Departments of Agriculture and Defense clung to the Dow theory for some weeks after it had been deflated. And the White House displayed no inclination to galvanize them into action.

In February 1970, Representative Richard McCarthy (D.-N.Y.), a leading opponent of the use of chemical and biological warfare techniques, wrote to the President's science advisor asking him why the government's commitment to restrict 2,4,5-T by January 1 had not been honored. DuBridge replied:

The October 29 announcement that you referred to was a statement of the actions that were planned to be taken by the various units of the Federal Government in relation to the 2,4,5-T. It was not a directive to the agencies for the simple reason that statutory responsibility for these decisions rests in the separate agencies.[33]

Representative McCarthy's reception of this explanation was understandably somewhat skeptical:

This is obviously a retreat from the position taken by the White House on October 29. As I read the statement at that time it was in the form of a directive that the departments will do such and such, now we find that the White House is backing off from this and is saying that the statutory authority rests with the agencies.

It seems to me that the President of the United States has authority—the ultimate authority over these agencies.[34]

On the same day (February 10) that Representative McCarthy received DuBridge's letter, Senator Philip Hart (D.-Mich.) announced that he would conduct hearings on the status of 2,4,5-T. Senator Hart's two days of hearings were held on April 7 and 15. This public exposure appears to have stirred the administration out of its paralysis once again. On the second day of hearings, Surgeon General Steinfeld began his testimony with the announcement that

new information reported to HEW on Monday, April 13, 1970, indicates that 2,4,5-T and its contaminant dioxins may produce abnormal development in unborn animals. Nearly pure 2,4,5-T was reported to cause birth defects when injected at high doses into experimental pregnant mice, but not in rats.[35]

Steinfeld was apparently trying to give the appearance of efficiency by saying that HEW had only learned of the teratogenicity of 2,4,5-T two days before. In fact, as we have already noted, these results had been reported at a meeting at the FDA (an agency within HEW) on February 24. (The rat experiment to which he referred was that by Dow, the experiment on mice by the National Institute of Environmental Health Sciences. Steinfeld did not mention an experiment done with hamsters at the FDA, which had also shown that purified 2, 4, 5-T causes birth defects.) It is also of interest that the government experiment which Steinfeld cited—that done by injection of mice with 2,4,5-T—was identical with the experiment done at Bionetics nearly four years before and labeled as being of uncertain significance by the government because (in Steinfeld's own words) "the 2,4,5-T had not been fed".

After announcing his "new information," Steinfeld proceeded to announce the restrictions which the government was imposing on the use of 2,4,5-T as a result: He announced

the immediate suspension by Agriculture of the registrations of the liquid formulations of the weed killer, 2,4,5-T, for use around the home and for registered uses on lakes, ponds, and ditch banks. . . . The Department of Agriculture intends to cancel registered uses of non-liquid-formulations of 2,4,5-T around the home and on all food crops for human consumption. . . for which it is presently registered. . . . These actions do not eliminate registered uses of 2,4,5-T for control of weed and brush on range, pasture, forests, rights of way and other non-agricultural land.[36]

The impact of this announcement was less dramatic than it might sound. The unaffected category of uses comprised about 75 percent of domestic usage of 2,4,5-T.[37] As for the "restrictions" on the remaining domestic uses, the public announcement did not make clear the significance of the distinction between the terms "suspension of registration" and "cancellation of registration."

Surely a majority of citizens hearing the announcement that the "registered uses of non-liquid formulations of 2,4,5-T around the home and on all food crops for human consumption" had been "canceled" would come to the conclusion that they need no longer worry about pregnant women being exposed to 2,4,5-T in their food or from weed killers applied to lawns. In fact, however, "cancellation" permits the use of pesticides until the chemical companies have exhausted a lengthy administrative appeal procedure. Only those few uses of 2,4,5-T for which the registration had been "suspended" were immediately affected, since "suspension" had the effect of outlawing these uses of the pesticide until the manufacturer could establish that they were safe. The choice between "suspension" and "cancellation" was made by the Agriculture Department according to whether or not, in its judgment, a use of 2,4,5-T was an "imminent hazard to the public."[38]

Another consequence of the administration's public disavowal of the Dow contaminant theory was that, on April 15, the Defense Department announced that Deputy Secretary of Defense David Packard had "temporarily suspended the use of 2,4,5-T for military operations pending further evaluation."[39]

The PSAC Review

One of the witnesses whom Senator Hart invited to appear at his hearings on the *Effects of 2,4,5-T on Man and the Environment* was the government official who had first made the Bionetics results public—Lee DuBridge, the President's science advisor. Instead of appearing in person, however, DuBridge sent a brief statement. The only new information which it contained was that, following his hurried announcement in October 1969 of government restrictions on the use of 2,4,5-T, DuBridge had appointed a panel of scientists under the President's Science Advisory Committee (PSAC) "to review all that is known about 2,4,5-T."[40] The statement continued: "This panel has prepared a report on the subject which I expect to make available within a few weeks."[41]

In fact, it was more than a year later before DuBridge's successor, Edward David, Jr., released the *Report on 2,4,5-T*—and then only after revelations by a group of independent scientists of the destruction resulting from the defoliation program in Vietnam had forced termination of the program in December 1970. The discussion in the PSAC report of the risks and benefits of domestic 2,4,5-T use seems reasonably objective—although critics have pointed out some crucial omissions.[42] The discussion of the use of 2,4,5-T in the South Vietnam defoliation program can only be characterized as a "whitewash."

The report discussed three aspects of the defoliation program: its military usefulness; the maximum possible amount of exposure of pregnant South Vietnamese women to 2,4,5-T and the possible teratogenic consequences of that exposure; and the ecological impact of the defoliation program.

The entire discussion of the military usefulness of the defoliation program was devoted to excerpts from testimony in which Rear Admiral W. E. Lemos had defended the program before a Congressional committee. The excerpts—which consist almost entirely of anecdotes concerning improvements in security in a few local areas as a result of the defoliation programs—seem almost irrelevant on the scale of justification required for a program which resulted in the defoliation of almost 10 percent of South Vietnam.[43] The report does not even mention the political impact in Vietnam of the defoliation program.

Regarding the possibility that use of 2,4,5-T had caused birth defects in Vietnam, the report dismissed what evidence there was with a sentence:

The lack of accurate epidemiological data on the incidence and kinds of birth defects in the Vietnamese population before or since the military use of defoliants precludes any estimate as to whether an increase in birth defects has occurred.[44]

The panel did not recommend that an attempt be made to collect such data. This initiative was taken later by independent scientists under the auspices of the American Association for the Advancement of Science. (See Chapter 11.) The panel then turned to theoretical "calculations of potential human exposures from sources such as drinking water or direct fall-out." From these calculations the panel concluded that the exposure of pregnant women to 2,4,5-T through their food or water could approach the levels at which birth defects had been caused in mice and rats. Each time it arrived at such a conclusion the panel quickly retreated, however, emphasizing how improbable it was for any individual to have suffered such an exposure. No mention was made of the possibility that birth defects in humans might be caused at lower levels of exposure than in rodents. (After the thalidomide disaster, it had been learned that the teratogenetic effect of equal proportions of thalidomide is 100 times greater on humans than on rats and 700 times greater than on hamsters.[45])

Finally, turning to the discussion in the report of the ecological impact of the defoliation program in South Vietnam, we find—nothing. Under the chapter heading "Some Ecological Effects" we find a listing of almost trivial items, such as that "when cottontail rabbits were given a choice of either 2,4,5-T treated vegetation or untreated, the rabbits consumed almost none of the treated vegetation"[46]; but we find not a single mention of the ecological impact of the defoliation and partial destruction of one-third of South Vietnam's jungle and the complete destruction of more than 20 percent of South Vietnam's mangrove forests by defoliation.

How can one account for the bias of the PSAC report on the subject of defoliation? One observer interviewed by the Washington correspondent of *Nature* magazine offered the explanation that "it was not the habit of PSAC to buck the Joint Chiefs of Staff, at least not under DuBridge."[47] Whatever the true explanation, the PSAC report on 2,4,5-T is further evidence of the decline of PSAC following the contemptuous treatment given its advice on the deployment of the Sentinel antiballistic missile system in 1967.

The Advisory Committee on the Chemical Companies' Appeal

The decision of the Agriculture Department to "cancel" rather than "suspend" the registration of 2,4,5-T for use on food crops was appealed by two of the manufacturers of 2,4,5-T, Dow Chemical and Hercules Corporation.[48] The appeal procedure required yet another advisory committee, appointed from a list of scientists provided by the National Academy of Sciences (NAS). (The NAS acted with apparent lack of concern for conflict of interest, including on its list of nominees one employee each of Dow Chemical and Monsanto, two of the three American chemical companies manufacturing 2,4,5-T.[49]) When the advisory committee finally reported its recommendations on May 7, 1971, it was not to the Secretary of Agriculture but instead to the Administrator of the newly created Environmental Protection Agency (EPA), which had taken over the responsibility for registering pesticides. The advisory committee report gave 2,4,5-T a clean bill of health—provided that the dioxin contamination was reduced to specified low levels.

One member of the advisory committee, Theodore Sterling, an Assistant Professor of Biostatistics at Washington University in St. Louis, disagreed and filed a minority report. Sterling agreed that it had not been established that 2,4,5-T was a public health hazard, but he also felt that it was premature to exonerate the chemical. He therefore concluded:

The Surgeon General was justified in feeling that a prudent course of action must be based on the decision that exposure to this herbicide may present an imminent hazard to women of child-bearing age. Hence, we [the advisory committee] can only recommend that the registration of 2,4,5-T be suspended and/or cancelled for use around the home, recreation areas, and similar sites and on all crops intended for human consumption. However, use of 2,4,5-T may be permitted under certain conditions for uses in forestation and rights of way.[50]

Sterling's dissent had no impact within the EPA. Staff scientists reviewed the report and appear to have endorsed the conclusions of the majority.

The EPA Advisory Report is Leaked

EPA Administrator William Ruckelshaus presumably would have implemented the advisory committee's recommendations in due course if the report had not been leaked to outside scientists, some of whom found themselves in much closer accord with Sterling's conclusions than with those of the committee's majority. On July 14, 1971, a group of these scientists organized by the Committee for Environmental Information and Ralph Nader's Center for the Study of Responsive Law held a news conference in Washington, D.C., in which they presented criticisms of the advisory report substantially the same as Sterling's.[51]

This time the EPA administration apparently heard the criticisms for it responded by turning for advice to scientists outside the agency—notably to scientists in the Food and Drug Administration who had conducted many of the experiments on the teratogenicity of 2,4,5-T. (It should be noted that, while the Agriculture Department-EPA advisory committee had not consulted these scientists, it *had* consulted with spokesmen for the manufacturers of 2,4,5-T. The advisory committee had even been presented with the results of a new study commissioned from the Bionetics Research Laboratories by one of the petitioners, the Hercules Corporation. This new study, represented as a replication of the original Bionetics study using purified 2,4,5-T, reported no birth defects. An investigation revealed an "error," however: in its "repeat study" Bionetics had used dosages of 2,4,5-T more than ten times smaller than those used in the original experiment.[52]) Following these consultations, Ruckelshaus decided to reject the advisory committee report and to go on to the next stage of the appeals procedure: public hearings.[53] At the time of this writing the hearings—after being delayed by a Dow Chemical Company lawsuit for two years[54] —are scheduled to begin in April 1974.

Thus we see how, more than ten years after Rachel Carson's first warning and five years after the first Bionetics report on the teratogenicity of 2,4,5-T, after the Mrak Commission report, the PSAC panel report, and the EPA advisory committee report, the government was still asking for advice as to what measures, if any, it should take to restrict 2,4,5-T. Meanwhile, the chemical companies continued to sell the chemical to whomever would buy it. It should also be noted that, although debate focused on 2,4,5-T, this chemical was only one of ten found to be teratogenic by Bionetics in the small sample of pesticides that it tested. Hence the title of our chapter.

NOTES

1. Rachel Carson, *Silent Spring* (Boston: Houghton Mifflin Co., 1962).

2. U.S., Executive Office of the President, Office of Science and Technology, *Use of Pesticides,* a Report of the President's Science Advisory Committee (Washington D.C.: Government Printing Office, May 1963). See also pp. 43-45 above.

3. Testimony of Dr. Jesse Steinfeld, Surgeon General, Department of Health, Education, and Welfare, before the U.S. Congress, Senate, Committee on Commerce, *Effects of 2,4,5-T on Man and the Environment* (hereafter referred to as *Effects of 2,4,5-T*), 91st Congress, 2nd Session, April 7 and 15, 1970. In this testimony will be found a chronology of government activities with regard to the determination of the teratogenicity of 2,4,5-T, along with Steinfeld's explanatory remarks (pp. 178-180).

4. Arthur Galston, "Warfare with Herbicides in Vietnam," in *Patient Earth,* edited by John Harte and Robert Socolow (New York: Holt, Rinehart and Winston, 1971), pp. 139-140.

5. U.S., Executive Office of the President, Office of Science and Technology, *Report on 2,4,5-T,* A report of the Panel on Herbicides of the President's Science Advisory Committee (Washington, D.C.: Government Printing Office, March 1971), p. 26.

6. Thomas Whiteside, *Defoliation* (New York: Ballantine, 1970), p. 1. Military Assistance Command figures for yearly total acreage defoliated and acreage of crop destruction in South Vietnam for 1962 through the first quarter of 1969 are given on p. 85. More complete figures are given in Chapter, reference 55.

7. *Report on 2,4,5-T,* p. 29.

8. Ibid. p. 26.

9. Steinfeld, *Effects of 2,4,5-T,* pp. 178-180.

10. Ibid.

11. Ibid.

12. Ibid.

13. Ibid.

14. The Bionetics results were first released in U.S., Department of Health, Education, and Welfare, *Report of the Secretary's Commission on Pesticides and Their Relationship to Environmental Health* (hereafter referred to as *Report of the Secretary's Commission on Pesticides*), (Washington, D.C.: Government Printing Office, December 1969), pp. 665-674. See also K. Diane Courtney, D. W. Gaylor, M. D. Hogan, H. L. Falk, R. B. Bates, and I. Mitchell, "Teratogenic Evaluation of 2,4,5-T," *Science* 168 (1970): 864.

15. James Singer, "DDT Debate Warms Up Again: Should the Government Restrict Its Use?" *National Journal,* 1 (1969): p. 1. John E. Blodgett of the Congressional Research Service in an unpublished manuscript, *Federal Ad Hoc Committees on Pesticides, 1955-1969* (July 1972), pp. II-11, II-12, describes in greater detail some of the pressures on Finch at this time. They included Government Accounting Office (GAO) reports, which described how the Department of Agriculture was ignoring HEW's advice about pesticide regulation, a Nader Summer Study Group investigation into the food protection activities of HEW's Food and Drug Administration, and, of course, the Bionetics Laboratories results.

Blodgett quotes one of Finch's aides as saying, "It was a political expedience sort of thing—Finch was being clobbered from all sides on pesticides safety."

16. Dr. Paul Kotin, Director of the National Institutes of Health, defended this refusal in his testimony in *Effects of 2,4,5-T,* pp. 94-97.

17. Quoted by Whiteside, *Defoliation,* p. 19.

18. *Report of the Secretary's Commission on Pesticides,* pp. 657-658.

19. Whiteside, *Defoliation,* p. 21.

20. Ralph Blumenthal, "U.S. Shows Signs of Concern Over Effect of 9-Year Defoliation Program in Vietnam," *New York Times,* March 15, 1970, p. 14, stated: "Vietnamese newspapers have been suspended for publishing articles about birth defects allegedly attributed to defoliants, and the public Health Ministry declines to provide any statistics on normal and abnormal births."

21. Quoted in Whiteside, *Defoliation,* pp. 21-22.

22. Ibid.

23. Bryce Nelson, private communication. See also Bryce Nelson, "Herbicide Order on 2,4,5-T Issued at Unusually High Level", *Science* 166 (1969): 977.

24. Quoted in Whiteside, *Defoliation,* pp. 94-95.

25. Ibid.

26. *Report of the Secretary's Commission on Pesticides.*

27. Testimony of Harrison Wellford, Center for Study of Responsive Law, Washington, D.C., in *Effects of 2,4,5-T,* pp. 7-8.

28. Department of Agriculture: Testimony of Dr. Ned D. Bayley, Director of Science and Education, Department of Agriculture, *Effects of 2,4,5-T,* p. 35. Department of Defense: inserted into the record, U.S. Congress, House, Committee on Armed Services, *Hearings on Research, Development, Testing, and Evaluation Program for Fiscal Year 1971* (hereafter referred to as *Hearings on 1971 RDT and E*) 91st Congress, 2nd Session, March 10-25, 1971, Part II, Appendix, pp. vii-viii.

29. Testimony of Dr. Julius E. Johnson, Vice President and Director of Research, Dow Chemical Company, *Effects of 2,4,5-T,* pp. 374-377.

30. Testimony of Surgeon General Steinfeld, *Effects of 2,4,5-T,* pp. 178-180.

31. Ibid.

32. Testimony of Dr. Samuel S. Epstein, co-chairman of the teratology panel of the Secretary's Commission on Pesticides, in *Effects of 2,4,5-T,* pp. 409-411. See also Nicholas Wade, "Dow Redefines Word It Doesn't Like," *Science* 176 (1972): 262.

33. DuBridge's letter and McCarthy's comments are reprinted in *Effects of 2,4,5-T,* p. 144.

34. Ibid.

35. The news release from which Dr. Steinfeld read these quotes is reprinted in *Effects of 2,4,5-T.*

36. Ibid.

37. Jamie Heard, "Restrictions on Controversial 2,4,5-T Fail to Satisfy Weed Killer's Critics", *National Journal,* April 25, 1970, p. 872.

38. Federal Insecticide, Fungicide and Rodenticide Act (7 U.S.C. 135) as amended in 1964. See Chapter 10 and the presentation of the legal battle over DDT for a more detailed discussion of these fine distinctions.

39. *Hearings on 1971 RDT and E,* Part II, Appendix, pp. vii-viii.

40. DuBridge's statement may be found reprinted in *Effects of 2,4,5-T,* pp. 452-455.

41. Ibid.

42. "PSAC Hiccoughs Over 2,4,5-T," *Nature,* May 28, 1971, p. 210.

43. *The Effects of Herbicides in South Vietnam: Summary and Conclusions* (Washington: National Academy of Sciences, 1974) p. 5-6.

44. *Report on 2,4,5-T,* p. 3.

45. Testimony of Samuel Epstein, M.D., U.S. Congress, Senate, Government Operations Committee, *Hearings on Chemicals and the Future,* 92nd Congress, 1st Session, April, 1971, p. 51.

46. *Report on 2,4,5-T,* p. 65.

47. "PSAC Hiccoughs Over 2,4,5-T," *Nature.*

48. Epstein, *Hearings on Chemicals and the Future.*

49. Nicholas Wade, "2,4,5-T Committee: Bias Untested, Academy Embarrassed", *Science* 173 (1971) p. 611.

50. U.S. Environmental Protection Agency, *Report of the Advisory Committee on 2,4,5-T,* May 7, 1971, p. 69.

51. Constance Holden, "Critics Weigh EPA Herbicide Report, Find it Wanting," *Science* 173 (1971) p. 312; see also: Committee for Environmental Information, "Critique of the Report of the Advisory Committee on 2,4,5-T," *Environment,* September 1971, p. 24.

52. Nicholas Wade, "Decision on 2,4,5-T: Leaked Reports Compel Regulatory Responsibility", *Science* 173 (1971) pp. 610-615.

53. Terri Aaronson, "Gamble", *Environment*, September 1971, p. 21.

54. Dow Chemical Co. v. Ruckelshaus, 477 F. 2d 1317 (C.A. 8th Cir. 1973).

CHAPTER 7

The Politician's Helper: Legitimizing the Cyclamates Decision

> *It is discouraging to find such conduct among public officials at the very time we are trying to impress upon our young people the importance of law and order.*
>
> —Representative L. H. Fountain
> on releasing the report of
> his subcommittee on federal
> regulation of cyclamate
> sweeteners.[1]

Advisory reports can be suppressed when their results are unwelcome or they can be commissioned as alternatives to facing up to unpleasant decisions, but at least the reports themselves are potentially useful if they get into the right hands—or are they? The case of the Medical Advisory Committee on Cyclamates illustrates dramatically that the advisory system itself can easily be corrupted. In this case, a government official who apparently wanted to give a political decision the appearance of technical legitimacy put together a committee of "experts" who obediently found reasons to tell him—and the public—what he wanted to hear.

Cyclamates were first used commercially as an artificial sweetener of foods in the early 1950s—primarily in special diets for the treatment of such conditions as diabetes. But in the 1960s their use became much more widespread, as the food industry conducted massive TV advertising campaigns extolling "diet" foods and soft drinks while panning over the contours of beautiful slim women.

On October 18, 1969, this commercial success story was suddenly jeopardized. Robert Finch, Secretary of Health, Education, and Welfare, called a press conference and announced:

I am today ordering that the artificial sweetener, cyclamate, be removed from the list of substances generally recognized as safe for use in foods.

Recent experiments conducted on laboratory animals disclosed the presence of malignant bladder tumors after these animals had been subjected to strong dose levels of cyclamates for long periods.

The findings of these experiments form the basis of my action.[2]

But Finch added that cyclamate-sweetened foods nevertheless would still be available.

My order does not require the total disappearance from the marketplace of soft drinks, foods, and nonprescription drugs containing cyclamates.

These products will continue to be available to persons whose health depends upon them, such as those under medical care for such conditions as diabetes and obesity.

I expect that in the future these products will be labeled as drugs to be consumed on the advice of a physician.[3]

The facts seem clear from the Secretary's statement: a new and unexpected danger had been discovered, and the government had moved decisively to protect the public from that danger. The government was just doing its job protecting the wholesomeness of the Nation's food supply. A look at the regulatory history of cyclamates both before and after Secretary Finch's announcement tells a much more complex story, however. In this chapter we investigate the role played by outside advisors in the process which (1) led the responsible federal agency, the Food and Drug Administration, to classify cyclamates as "Generally Recognized as Safe" during the period 1958-1969; (2) prompted Secretary Finch to conclude in 1969 that the benefits to "persons whose health depends upon them" outweighed the risks; and (3) led the Department of Health, Education, and Welfare (HEW) to reverse this decision a year later, finally banning cyclamates entirely after most of the cyclamate-sweetened food already on store shelves and in warehouses in October 1969 had been sold.

The Food and Drug Administration and the National Academy of Sciences

The use of cyclamates as a food additive became established in an era when such chemicals were given the benefit of the doubt. In the early fifties the burden was on the Food and Drug Administration (FDA) to prove that food additives were *un*safe in order to force their withdrawal from use. But the agency was not looking for fights with the food industry. Unless there were blatant adverse health effects from a food additive, the FDA was inclined to look the other way. This is what happened with cyclamates.

In 1958, with passage of the Food Additives Amendment to the Federal Food, Drug, and Cosmetics Act, manufacturers of food additives were required to prove to the FDA that their products were safe—unless a food additive was

generally recognized, among experts qualified by scientific training and experience to evaluate its safety, as having been adequately shown through scientific procedures (or in the case of a substance used in food prior to January 1, 1958, through either scientific procedures or experience based on common use in foods) to be safe under the conditions of its intended use.[4]

This exemption led to the compilation by the FDA of a "Generally Recognized as Safe" (GRAS) list of food additives.

The advice that the FDA had received from the Food and Nutrition Board of the National Academy of Sciences' National Research Council (NAS-NRC) would not appear to imply that cyclamates were generally recognized as safe. The Board's 1954 advisory report concludes:

The Board is impressed with the fact that cyclamate has physiologic activity in addition to its sweetening effect, that there is no prolonged experience with its use, and that little is known of the results of its continued ingestion in large amounts in a variety of situations in individuals of all ages and states of health. The priority of public welfare over all other considerations precludes, therefore, the uncontrolled distribution of foodstuffs containing cyclamate.[5]

But the FDA decided that a careful look at the health effects of cyclamates was not required and included cyclamates on the "Generally Recognized as Safe" (GRAS) list along with several hundred other food additives and common household seasonings.

The food industry had a strong economic incentive to maximize its use of cyclamates: cyclamates provide sweetening power at about one-tenth the price of sugar, and the label "diet drink" or "diet food" had obvious appeal to weight-conscious Americans. The FDA's action in placing cyclamates in the same category of safety as sugar, salt, and cornstarch was understood by the industry as permission to go full speed ahead. The advertising men were unleashed, and national consumption of cyclamates skyrocketed from about 1 million pounds in 1958 to about 17 million pounds in 1968.[6]

The FDA was somewhat taken aback by this tremendous increase in the use of cyclamates. In 1962 the NAS-NRC Food and Nutrition Board was asked to look once again into the safety of cyclamates. The conclusion of its report was the same as before.

The priority of public welfare over all other considerations precludes, therefore, the uncontrolled distribution of foodstuffs containing cyclamate.[7]

The report added:

It is emphasized strongly that the availability and consumption of artificially sweetened foodstuffs have no direct influence on body weight, nor are the foodstuffs in question of any importance in weight reducing programs except as they are used in feeding regimens in which the total energy intake is supervised and controlled.[8]

This statement reflected evidence that cyclamates may actually be an appetite stimulant[9] and, of course, directly contradicted the claims then being made in the massive advertising campaigns promoting the consumption of cyclamate-sweetened foods and drinks.

Although the new NAS-NRC report did not cause the FDA to remove cyclamates from the GRAS list, it has been credited with stimulating research into the possible adverse effects of cyclamates.[10] As the 1960s went on, this research turned up increasing evidence for a long list of serious side effects associated with cyclamates use, ranging from major changes in the actions of drugs in the presence of cyclamates to growth retardation, liver damage, chromosomal damage, and birth defects.[11]

In 1968 the FDA repeated its ritual of asking the NAS-NRC for a review of the safety of cyclamates. Once again the ritual response came back that "totally unrestricted use of the cyclamates is not warranted at this time."[12] It was now fourteen years since the FDA had first received this warning, and the scientific evidence for adverse effects had mounted to the extent where there was considerable concern about cyclamates in the medical and scientific divisions of the FDA. Congressional staffers investigating in 1970 turned up a number of internal memoranda dating from late 1968 urging higher-ups to take cyclamates off of the GRAS list.[13] Foods and drinks containing cyclamates had become a billion-dollar-a-year business,[14] however, and the FDA brass apparently relished less than ever the prospect of the bruising confrontation with industry which would have developed if an attempt had been made at that time to remove cyclamates from the GRAS list. As one internal FDA memorandum stated in September 1967:

We cannot say today that the cyclamates are generally recognized as safe; however, removing them from the GRAS List and establishing tolerances in soft drinks, *et cetera*, will produce difficult problems.[15]

The Congressional subcommittee which in 1970 investigated the handling of the cyclamates affair summarized the situation as it stood before October 1969 as follows:

It was evident at least as early as 1966 that there was a genuine difference of opinion among qualified experts as to the safety of cyclamate sweeteners. Consequently, FDA had an obligation at that time to remove cyclamates from the GRAS List, to declare them to be a "food additive" within the statutory definition, and to ban their use until industry had established their safety. But despite the mounting evidence in the ensuing years, FDA did not act.[16]

The Sugar Research Foundation and the
Delaney Amendment

Action was finally forced in October 1969 by an initiative from within the food industry itself. The sugar industry had not enjoyed seeing cyclamates taking over its market and had funded research on the side effects of this food additive through its Sugar Research Foundation. The research eventually led to the conclusion that cyclamates produce bladder cancer in rats.[17] This discovery activated a section of the Food Additives Amendment, the "Delaney Clause," which specifies

that no additive shall be deemed to be safe if it is found to produce cancer when ingested by man or animal or if it is found, after tests which are appropriate for the evaluation of the safety of food additives, to induce cancer in man or animal.[18]

In other words the FDA now had no choice but to ban cyclamates as a food additive.

Thus followed the October 1969 announcement made by HEW Secretary Finch (within whose Department the FDA resides). Finch, a long-term political associate of President Nixon, anticipated the cries of anguish from the food industry and did the best he could to soften the blow. He promised that cyclamate-sweetened foods and drinks could continue to be sold if they were relabeled as "nonprescription drugs" and moved to appropriate supermarket shelves. He also promptly accepted the suggestion by the fruit canning industry, which had just completed its canning season in his home state of California, that the deadline for removing foods containing cyclamates from the market be postponed seven months.[19] He even went so far as to initiate efforts to repeal the Delaney clause.[20]

Secretary Finch's Medical Advisory Committee on
Cyclamates

Having publicly promised that cyclamate-sweetened foods would remain available as nonprescription drugs, Finch found himself in an uncomfortable position. The FDA—which was legally responsible for the registration of new drugs—pointed out that registering these products as drugs would probably be illegal, for drugs are required by law to have been shown by their manufacturer to be safe and to be effective against some disease. But in the words of an internal FDA position paper:

We are aware of no evidence that cyclamate-containing foods are safe or effective in the treatment of obesity or diabetes. Under the principles we strongly adhere to in permitting drugs to be marketed, these products should not be allowed on the market. To approve a New Drug Application for these products is not supportable medically or legally.[21]

Finch was committed, however. If he couldn't get the FDA's blessing, then he would find other experts. The Secretary's Medical Advisory Committee on Cyclamates was duly set up, made up in almost equal numbers of HEW administrators (Finch's subordinates) and outside specialists.[22] And after due consideration of the evidence submitted to it by the FDA, the committee gave Finch the advice he wanted:

Although the use of cyclamates is not absolutely necessary in any disease, it can be useful in the medical management of individuals with diabetes or patients in whom weight reduction and control are essential to health. Particularly in juvenile patients who have diabetes, where sweets and soft drinks are a special problem, non-nutritive sweetened foods may be an essential part of preventative therapy.[23]

The advisory committee also gratuitously informed the Secretary of their support on another point. They advised that foods and drinks containing cyclamates remain available "on a non-prescription drug-labeled basis to be used only on the advice of a physician."[24]

Unfortunately for the committee—and the Secretary—this recommendation was to cause trouble. Not only did it violate common sense to put a medicine which was "to be used only on the advice of a physician" into the category of nonprescription drugs, it also violated the specific requirements of the Federal Food, Drug, and Cosmetics Act which defines a prescription drug as

a drug intended for use by man which because of its toxicity or other potentiality for harmful effect, other methods of its use, or other collateral measures necessary for its use, is not safe for use except under the practitioner licensed by law to administer such drug.[25]

It is puzzling why a committee made up entirely of M.D.s took a position that it must have known was indefensible. The only obvious advantage from such a recommendation would accrue to the distributors of cyclamate-sweetened foodstuffs, who would be able to continue to deal with their customary grocery store outlets.[26] But most people would agree that such considerations are outside the province of a medical advisory committee.

The Congressional Investigation

If it had not been for a group of "Nader's raiders," the story might have ended here. In early 1970, a report of a Ralph Nader summer study group on the FDA was released: *The Chemical Feast* by James S. Turner. A study of the background of Secretary Finch's cyclamates decision was the book's featured attraction—Chapter One. The discussion quoted extensively from the FDA's files and was based also on interviews with FDA personnel. An excerpt will indicate the message:

The dramatic removal of cyclamates from the marketplace was necessary because the FDA failed to do its job. It did not heed the frequent early warnings against the general use of cyclamates made by the scientific community. It did not periodically and systematically review the safety of substances on its GRAS list. It dismissed or distorted the warnings of its own scientists. Secretary Finch compounded these failures by ignoring the accumulated doubts about cyclamates and minimizing the importance of removing the chemical from the market rapidly. He did not connect this removal with the legal requirement that all chemicals must be proved safe before being added to food. He never mentioned evidence that birth defects and genetic damage that were related to cyclamates in tests on laboratory animals are a more serious danger than cancer. And he denied the importance of free scientific inquiry, expression, and interchange between scientists and the public. . . . By attempting to avoid, then delaying and finally distorting the ban on cyclamates, the FDA and Secretary Finch undermined confidence in the American Food supply and left the impression that neither government nor industry is primarily concerned with protecting the public interest.

The impression is quite accurate.[27]

The charges contained in *The Chemical Feast* helped bring about a Congressional investigation of the FDA's handling of the cyclamates affair by the Intergovernmental Relations Subcommittee of the House Committee on Government Operations. The subcommittee is headed by conservative North Carolina Democratic Representative L. H. Fountain. The staff of Fountain's subcommittee did a thorough study of the FDA's records relating to the matter and explored a number of aspects of the affair which the Nader report had failed to develop—the role of the Secretary's Medical Advisory Committee on Cyclamates in particular.

When newly appointed FDA Commissioner Edwards came before the subcommittee, Congressman Fountain did not mince words:

I believe that this subcommittee can, within the limitations of time and staff, render a public service in reminding you, Dr. Edwards, and your associates, that the role of FDA is to enforce the [Food, Drug, and Cosmetic] act fully and effectively. All of the sections of the law are important, and Congress did not, and I believe does not now want any of them to be put in limbo, as I am sure some people would like.[28]

This opening statement was then followed by relentless questioning of Edwards and his subordinate administrators by Fountain and two members of his subcommittee staff, Gilbert Goldhammer and Dr. Delphis Goldberg. Memorandum after memorandum from the FDA files and addressed by FDA's medical and scientific staff to its administration were introduced. In these memoranda the adverse health effects of cyclamates were repeatedly set forth as a basis for removing cyclamates from the GRAS list. As the documents piled up, Edwards and his staff offered an ever weaker defense of the FDA's record, until finally Fountain squeezed this admission from Edwards:

I think without any question the cyclamates could have been removed from the GRAS list earlier than they were. I am not prepared, Mr. Chairman, to say specifically when, but I think it could have been done considerably sooner than it was.[29]

The FDA officials *did* defend Finch's decision to relabel cyclamate-sweetened foods as nonprescription drugs, but Fountain's subcommittee was not persuaded. In a report to the House based on the hearing record, Finch's role in the cyclamates affair was described as follows:

The Secretary of Health, Education, and Welfare announced on October 18, 1969, that prohibition of further marketing of cyclamate-containing products as foods was required by the Delaney Clause. The Secretary announced at the same time that continued marketing of cyclamate-containing products as non-prescription drugs would be permitted. FDA was then called upon to implement this decision, which the agency sought to do through *illegal regulations and procedures.* The basic cyclamate decisions were made in the Secretary's office despite the fact that responsibility for enforcing the Federal Food, Drug, and Cosmetic Act had been delegated to the FDA Commissioner. [Emphasis added.][30]

The Medical Advisory Committee Meets Again

Just before the June 1970 Congressional hearing, Finch was replaced as HEW Secretary and appointed Counselor to President Nixon, a position in which he quickly faded into well-deserved obscurity. And with Finch out of the way, HEW moved to extract itself from its increasingly untenable position on cyclamates. The way in which this was done was true to form. HEW reconvened the Medical Advisory Committee and asked it to reconsider the safety and effectiveness of cyclamates. The response to this request was dramatic to say the least: the committee reversed itself completely. It explained its change of mind by citing "new information" on the production of bladder tumors in rats with doses of cyclamates comparable (relative to body weight) to those consumed by heavy cyclamate users. The committee added that

the literature provided to the group does not contain acceptable evidence that cyclamate has been demonstrated to be efficacious in the treatment or control of diabetes or obesity.[31]

The committee offered no explanation for this direct contradiction of its previous assertion that cyclamate-sweetened foods may be an essential part of preventative therapy with juvenile diabetics. Cyclamates were thereupon totally banned.

Representative Fountain was not through, however. His subcommittee staff investigated the matter again and established that the evidence which had been cited by the Medical Advisory Committee as "new" had in fact already been referred to in its original report.[32] Indeed, little had changed in the interval between the Committee's two meetings other than the political pressures on HEW generated by the Fountain subcommittee hearings. The subcommittee's final report did not conceal its disdain at the way in which HEW had used its Medical Advisory Committee:

HEW used an outside advisory body to make recommendations on matters which had already been decided, involving a basic issue which the advisory body was not qualified to decide.

At the time HEW convened the medical advisory group on cyclamates, the Secretary had already announced publicly that cyclamate sweeteners and cyclamate-containing food products would be available in the future as non-prescription drugs. In affirming the Secretary's decision, the group acted on the same scientific facts that had been considered by FDA's medical staff in reaching a contrary conclusion. The advisory group, moreover, was not qualified to determine the real issue—whether the law permitted implementation of the Department's announced decision to permit continued marketing of cyclamates.

Similarly, the reconvening of the Medical Advisory group served no valid scientific purpose after the subcommittee's hearings had spotlighted FDA's illegal cyclamate regulations. The evidence on which the panel reversed its earlier recommendations was known and available to the group when it was originally convened.[33]

HEW responded in kind by issuing a press release which claimed to rebut the Congressional report and concluded by stating that "its [the subcommittee report's] interpretation of the facts and the law in this instance are erroneous."[34] When Fountain requested Dr. Edwards to explain in person to the subcommittee the error in its interpretation, however, Edwards put on a rather pathetic performance.[35]

The prostitution of the advisory committee system in this case is obvious and needs no further comment. Another point worth noting, however, is the remarkable ineffectuality of the NAS-NRC Food and Nutrition Board in its fourteen years of advising on the cyclamates issue. It makes one wonder why such advisors keep coming quietly back.

NOTES

1. Quoted by R. J. Bazell, "Cyclamates: House Report Charges Administrative Alchemy at HEW," *Science* 170 (1970): 419.

2. Quoted in U.S. Congress, House, Committee on Government Operations, *Cyclamate Sweeteners,* 91st Congress, 2nd Session, June 10, 1970. p. 27.

3. *Ibid.*

4. Quoted in U.S. Congress, House, Committee on Government Operations Report, *Regulation of Cyclamate Sweeteners,* 91st Congress, 2nd Session, House Report No. 91-1585, October 8, 1970, p. 3.

5. National Academy of Sciences-National Research Council Food and Nutrition Board, *Policy Statement on Artificial Sweeteners* (Washington, D.C.: National Academy of Sciences, November, 1954), pp. 7-8.

6. U.S. Department of Agriculture figures, reprinted in *Cyclamate Sweeteners,* p. 63. The price relative to sugar is quoted on p. 6.

7. NAS-NRC Food and Nutrition Board, *Policy Statement on Artificial Sweeteners* (Washington, D.C.: National Academy of Sciences, revised April 1962), p. 7.

8. *Ibid.*, p. 4.

9. A study by M. B. McCann, *et al., Journal of the American Dietetic Association* 32 (1956): 327, found "no significant difference . . . when the weight loss of users and nonusers of these products [artificial sweeteners] was compared." A later controlled experiment with rats by L. M. Dalderup and W. Visser, "Effects of Sodium Cyclamate on the Growth of Rats Compared With Other Variations in the Diet", *Nature* 221 (1969): 91, found that "sodium cyclamates in the quantity given [replacement for sugar in ordinary diet] seems to stimulate appetite, and thus a rise in weight. It is surprising that the food efficiency [weight gained per calorie consumed] is also higher." Both articles are quoted in *Cyclamates Sweeteners,* pp. 64-68.

10. Report by John J. Schrogie, M.D., October 1969, reprinted in *Cyclamate Sweeteners,* pp. 4-5.

11. See e.g., the FDA reports reproduced in *Cyclamate Sweeteners,* pp. 18-21 and 82-84.

12. Quoted in FDA memorandum reproduced in *Cyclamate Sweeteners,* pp. 18-21.

13. Quoted in *ibid.*, pp. 17-21.

14. *New York Times,* October 19, 1961, p. 1.

15. Quoted in *Cyclamate Sweeteners,* p. 22.

16. *Regulation of Cyclamate Sweeteners,* p. 5.

17. The sequence of events which led to this conclusion is recounted in a letter to Representative L. H. Fountain (D.-N.C.) from R. W. Kasperson, vice-president of Abbot Laboratories, a manufacturer of cyclamates. The latter is reproduced in *Cyclamate Sweeteners,* pp. 98-103.

18. Quoted in *Regulation of Cyclamate Sweeteners,* p. 4.

19. James S. Turner, *The Chemical Feast: Ralph Nader Study Group Report on Food Protection and the Food and Drug Administration* (New York: Grossman Publishers, 1970) p. 19. See also *New York Times,* October 23, 1969, p. 36.

20. See the story by G. C. Thelen, Jr., *Washington Post,* June 10, 1970, reproduced in *Cyclamate Sweeteners,* pp. 35-36, and the discussion there (pp. 34-36) between Representative Fountain and FDA Commissioner Edwards concerning the allegations in the article. See also *New York Times,* November 6, 1969, p. 5.

21. Quoted in *Cyclamate Sweeteners,* pp. 68-69. See also the discussion following p. 69.

22. The membership of the Medical Advisory Committee is given in *ibid.*, p. 40.

23. Quoted in *ibid.*, p. 89. The entire first report of the Medical Advisory Committee is reprinted in *ibid.*, pp. 86-90.

24. Quoted in *ibid.*, p. 87.

25. Federal Food, Drug, and Cosmetics Act, Section 503(b)(1)(B) (21 U.S.C. 353(b)(1)(B)). Quoted in *ibid.*, p. 91.

26. See the discussion on this point between Representative Fountain and FDA spokesmen quoted in *ibid.*, pp. 91-93.

27. Turner, *The Chemical Feast,* p. 29.

28. *Cyclamate Sweeteners,* p. 1.

29. *Ibid.*,p. 24.

30. *Regulation of Cyclamate Sweeteners,* p. 15.

31. The second report of the Medical Advisory Committee on Cyclamates is reprinted in U.S. Congress, House, Committee on Government Operations, *The Safety and Effectiveness of New Drugs (Market Withdrawal of Drugs Containing Cyclamates)*, 92nd Congress, 1st Session, May 3, 1971, pp. 20-27. This reference is hereafter referred to as *Safety and Effectiveness of New Drugs*. The lines quoted appear on p. 21.

32. *Regulation of Cyclamate Sweeteners*, pp. 12-13.

33. *Ibid.*, p. 16.

34. U.S. Department of Health, Education, and Welfare, *HEW Position Paper*, reprinted in *Safety and Effectiveness of New Drugs*, p. 17.

35. *Ibid.*, p. 15ff.

PART III

Responsibilities of Science Advisors in a Democracy

The argument given ... is that the President is entitled to the best scientific advice available, and that any advice given to him is a personal service which must remain confidential. This might be persuasive if the President were king. However, in our society there is a higher priority: it is that the citizen must have the best scientific advice.

—Charles Schwartz
(in *The Nation*,
June 22, 1970)

CHAPTER 8

The Advisor's Dilemma

I have a feeling that a lot of them see me with a kind of horror—not just anger, but with an awe of the sort you'd have for an astronaut who stepped out of that capsule and cut his umbilical cord and just floated off into space and had become weightless, drifting in a black void, because he cut himself off from the capsule and from NASA, and the U.S. government, and the U.S. budget that supports that entire system. . . .

I think four-year-olds have fantasies like that . . . of what the world would be like when the mother went away. And the mother is the U.S. Executive Branch.[1]

—Daniel Ellsberg describing the
reactions of his colleagues
at the Rand Corporation
after he made public the
"Pentagon Papers"

The executive branch's science advisory establishment makes many essential contributions to the effectiveness of policy making. It is also obvious from our case studies, however, that administration officials have learned to use the advisory establishment to mislead the public and Congress about the technical bases of executive decisions. In any particular case the advisor must therefore decide whether he is being asked to advise or to "legitimize." But what then? If he refuses to participate in a system which is being used to mislead the public, he will also be refusing to give his government the benefit of his advice. Such is the advisor's dilemma.

101

One deceptively easy resolution of this dilemma would appear to be for an advisor to say to himself: "I will give the administration the best advice that I can concerning technical considerations. Then, If I find that executive spokesmen start misleading the public about these considerations, I will give the public directly the benefit of my knowledge and experience."

Unfortunately, things are not quite so simple, because executive officials do not in general find such behavior acceptable. Advisors, like permanent government employees, are expected to be loyal and to abide quietly by final executive decisions, or else to "get off the team."

When an advisor decides to "go public" he is aware that he may very well at the same time be sacrificing his future access to the corridors of power and the sources of inside information. Since there are, in the first place, few advisors willing forcefully to present an unwelcome point of view to important government decision makers, an advisor can legitimately be concerned that his replacement by a "yes man" may in the long run outweigh any benefit the public might derive from his setting the public record straight on a particular issue. When concern about loss of future effectiveness within the executive branch is combined with the considerable doubt that most advisors have about the effectiveness of speaking out, it is not surprising that it is so extraordinarily rare that advisors "go public."

There are also strong social and psychological pressures operating against "going public." The high-level government advisor has typically undergone a long process of "socialization" in Washington during his slow climb up through the hierarchy of advisory committees. His self-esteem, not to mention his position in his organization and in the eyes of his colleagues, may not be unrelated to his advisory activities and his association with men in power.[2]

It is becoming more and more clear, however, that to the extent that the administration can succeed in keeping unfavorable information quiet and the public confused, the public welfare can be sacrificed with impunity to bureaucratic convenience and private gain. Thus advisors who keep their information and analyses confidential in the interests of preserving their "effectiveness" may find that very effectiveness decreasing as a poorly informed and uncertain Congress and public become less and less able to call the administration to account for irresponsible actions.

There is no consensus within the scientific community as to how the advisor's dilemma should be resolved. In fact, there has been very little discussion at all within the scientific community of the issues involved. Lack of such discussion leaves scientists unprepared when they become advisors and find themselves confronted with difficult and unfamiliar decisions, often in an atmosphere of great pressure. It is no wonder that under these circumstances advisors find themselves looking for guidance to the experienced government officials whom they advise and adopting rather uncritically the code of confidentiality and team spirit to which these officials themselves adhere.

Arguments Supporting the Confidential
Advisory Relationship

Let us consider a few of the arguments which, by and large, the advisors adopt as their own:

1. *The relationship between a scientific advisor and the government official whom he advises should be confidential, just as is that between a lawyer and his client.*

This analogy compares a scientist or engineer who provides information and advice to the government—presumably with the intention of helping bring forth an optimal policy for the country as a whole—with the private lawyer hired to devise the optimal strategy in presenting his client's case. If we follow this analogy through, it would appear that the executive branch sees itself in an adversary relationship with Congress and the public. The fact that one side in the confrontation has a near-monopoly on the "lawyers" (science advisors) then becomes quite disquieting.

It is unfortunate that the ethical principles proposed for advisors by executive-branch agencies have more in common with the ethics of lawyers and physicians, which stress the protection of the client, than with the ethics of responsible public officials or public health officers, for whom the general welfare must be the primary concern. Science advisors, who are concerned with questions of the national interest, should also owe their first loyalty to the nation as a whole and to fundamental democratic principles, rather than to the personalities or policies of any particular administration. Patterning the ethics of science advisors on those of private lawyers or physicians is therefore inappropriate.

2. *The President is elected by all the people and has the ultimate responsibility for making national policy. This leaves the advisor with only the responsibility of seeing that the President and the officials in his administration are well informed.*

In response to the great inequality of activity and influence which has developed among the three branches of our government, the popular identification of our form of government as democratic has come to depend less on the theory of checks and balances and more on the fact that the President is elected "by all the people." We might thus caricature this view of our government as the "Four-Year Elected-Dictatorship Theory of Democracy." This theory has been particularly popular with the Nixon administration, whose behavior has given the country a most vivid demonstration of the dangers posed by an executive branch which feels that it can be held to account only once every four years.

What the elected-dictatorship idea leaves out entirely is the role of the individual citizen in the governmental process. The ultimate responsibility under a democratic government always lies with the individual citizen, and the government advisor cannot escape his responsibilities as a citizen. In fact, by virtue of his greater knowledge of the subject on which he advises, the government advisor takes on enlarged responsibilities for the defense of the public interest in that area. The confusion of allegiance to the public interest with allegiance to the President in power indicates a basic lack of understanding of the meaning of democracy. That this misunderstanding has been shared by so many science advisors should be a matter of great concern to the scientific community as a whole.

Such concerns were raised about the long acquiescence of science advisors in Presidential policies for the Indochina War. Although a number of prominent scientists may have had private qualms about American actions in Vietnam, they confined themselves to producing a secret report, prepared during summer 1966 under the auspices of the "Jason" division of the Institute for Defense Analyses (IDA). The report argued against the bombing of North Vietnam, not on any moral grounds, but on the technical grounds of its ineffectiveness.[3] Their criticism of the bombing was largely ignored by the generals—although it appears to have influenced Defense Secretary McNamara, who attached its conclusions to a memorandum to President Johnson opposing the increased bombing of North Vietnam.[4] McNamara failed to convince Johnson and subsequently left the Pentagon. But a related proposal endorsed by the advisors was partially adopted: an electronically policed barrier along the norther borders of South Vietnam. The advisors claimed that such a barrier would be more effective than bombing in choking off the flow of military support to the Vietcong.[5] The result was the "McNamara Line," which ultimately grew into the military fantasy-nightmare of the "electronic battlefield."[6] But the bombing went on. One of the leaders of the Jason summer study told us that he was so embittered by this experience that he subsequently resigned from all his government advisory posts. "I was a dupe," he said. "Whatever advice you give the military will be twisted."

When government officials repeatedly fail to hear or heed their science advisors and when an advisory committee begins to moderate or even alter what it would really like to say (Trojan Horse strategy), advisors should perhaps consider other approaches. Bringing serious matters into the open and to the attention of government decision makers through their morning newspapers is one tactic for breaking through their bureaucracy-created isolation. It has been established repeatedly that public exposure of important issues can result in crucial facts and perceptions coming to the fore which would have been missed in the ordinary governmental process.

3.   *It is quixotic for a lone scientist with no political constituency to hope to influence the public to reject the misrepresentations of administration spokesmen.*

The case studies of outside activities to be presented in Part IV show that a lone scientist *can* fight the bureaucracy—and win. It is true that it is usually ineffective for an insider just to sign a petition or make a single public statement and then go back to his usual activities. This will probably only succeed in antagonizing those administration officials he has been advising. If an advisor wants to challenge an administration policy that he considers a threat to the public health and welfare, then his dedication in raising an opposition must be commensurate with the seriousness of the perceived threat. Great persistence and resourcefulness are also usually required—and often courage, too, since the scientist may be opposing agencies which fund his work or work at his institution.

Although serving as an advisor broadens one's first-hand knowledge of the considerations which enter into federal policy making for technology, it does not prepare one for the rigors of such a battle. Advisors are not encouraged to follow through on their advice and try to see that it is taken into account. Generally they are asked to prepare and submit reports rather quickly and then to forget about them unless called upon for further advice. Often, they are not expected to look seriously into the nontechnical aspects of the issue on which their advice is sought. Instead they are expected to form an opinion based primarily on the knowledge they already have and on the briefings they receive from government officials and from full-time government experts. They are paid for this, they gain prestige because their advice is sought by important government officials, and they make professional contacts which may prove important in the advancement of their careers. This is quite a different situation from the harsh and lonely world in which an independent scientist often finds himself.

Thus, of the three rationales offered in defense of the confidential advisory relationship, two—the lawyer-client analogy and the the-President-has-the-ultimate-responsibility argument—seem upon reflection to be absurd. The third, the you-can't-fight-city-hall argument is, as we said, simply a restatement of the fact that the life of a confidential advisor can be relatively easy and secure while that of the public interest scientist can be arduous and uncertain. As Abraham Lincoln said, "Silence makes men cowards."

It is obvious, from the superficiality of the widely held views which we have been discussing, that the ethics of advising should be subjected to a careful examination by the scientific community as a whole. Science advising, no less than scientific research, needs a code of ethics. And this code should explicitly take into account the fact that we live in a democracy in which the ultimate responsibility resides not with the President but with the individual citizen.

Discounting Future Effectiveness

The rather old-fashioned lecture on citizenship which we have just delivered does not by any means resolve the deeper dilemma in which a science advisor often finds himself: it simply acts to blow away the smoke screen concealing it.

Generalizations cannot resolve such dilemmas, for each case concerns an individual scientist's judgment of how he may most effectively serve the public interest. An advisor contemplating going public in order to challenge an emerging executive policy that he considers inimical to the public interest is weighing two great uncertainties: the effectiveness of such a move versus his future effectiveness as an insider if he maintains confidentiality.

The high-level advisor finds himself in a position which has usually required years of apprenticeship to arrive at. It is therefore natural, before challenging a policy, for him to think: "I've worked hard to gain my position of influence—for what it's worth. Let someone else take the issue to the public. That way I can keep presenting my arguments on the inside while they present theirs on the outside. (Besides, I'm the director of a large laboratory, and a lot more people will be hurt if I become unpopular with the current administration.)"

The problem, of course, is that such advisors represent a considerable segment of the leadership of science, and if they, in their positions of relative security, are unwilling even occasionally to set an example by taking the risk of going public, it is unreasonable to expect that enough high-caliber scientists outside the advisory establishment will step forward in their stead. Also, by asking other scientists to assume the entire burden of public interest science, the advisors may be asking them to close to themselves the doors to positions of honor and influence which the advisors themselves enjoy.

Unfortunately, it appears characteristic of human nature to overestimate what one's future effectiveness might be in comparison to what one judges one's effectiveness to be in the issue at hand. Participants in politics often must revise their hopes for future accomplishment down by an order of magnitude during the battle when they realize how tough it is to accomplish anything. This means that an advisor weighing the effectiveness of going public in a current situation is weighing this reduced expectation against his still-high hopes of future effectiveness. This gives rise to the apparently common situation where an advisor conserves his effectiveness like a beautiful girl her virginity—until no one is interested in it anymore.

What Does the Advisor Do About Uncertainty?

Uncertainty arising from incomplete information is one of the major problems facing a technical expert—advisor or not—when he is contemplating making an issue out of his concerns. Thus, taking examples from our case studies so far: It was not clear to what extent a fleet of SSTs would increase the earth's cloud cover or deplete its protective layer of stratospheric ozone. Nor was it clear how many birth defects would occur in South Vietnam from the massive use there of 2,4,5-T as a defoliant. And finally, it was not clear how many cases of cancer and birth defects would result from the public's massive use of cyclamate-sweetened drinks and foods.

A concerned scientist might therefore well have asked himself: "Is this a false alarm? Am I putting my reputation on the line over a danger which later information will prove not to exist?"

In these circumstances the decision must hinge on the advisor's answer to the question: Who should determine whether the benefits of the proposed policy exceed the risks? One PSAC panel, reporting on the safety of underground nuclear weapons testing, suggested that

the public should not be asked to accept risks resulting from purely internal governmental decisions if, without endangering national security, the information can be made public and the decisions reached after public discussion.[7]

(The panel's report was subsequently suppressed.) Thus, even if the dangers which concern a scientist might not materialize, the public should have an opportunity to express its opinion as to whether the potential risks are worth the benefits.

This does not mean that every such matter should be made the subject of a national referendum. What it does mean is that, in a democracy, the citizens should have an opportunity to defend their vital interests. Not infrequently an administration decision is made in secret and then, when the story gets out, the decision is reversed. What has happened is that the publicity has brought new political forces into play.

Guidelines for Advisors

While there are many cases in which advisors have refused to come forward to warn the public, we are unaware of a single case in which an advisor has sought publicity for an unfounded concern for the public welfare. It is not surprising that a bias exists toward acquiesing to the executive branch's demands for confidentiality: the advisors have working relationships with executive officials. It happens also that the counsels of timidity and ambition work in the same direction: no one has risen to high position by appealing over the government's head to the public, while many have constructed distinguished careers by playing the game according to administration rules.

Some advisors have not only accepted confidentiality as a necessity, they have even embraced it. Thus the technical society of operations analysts, the Operations Research Society of America (ORSA), includes in its "Guidelines for Professional Practice" the following admonitions:

Scrupulously observe any ground rules about confidentiality laid down by the organization being served.

Report the study's results only to the organizational elements sponsoring the study, unless specifically authorized by them to report to a wider audience.[8]

They further declare that

an analyst called upon to testify on behalf of a client whose decision he has helped to shape by his analyses should support his client's case. . . . An analyst who wishes to disagree publicly with his client is placed in a difficult ethical position.[9]

The perspective implicit in these guidelines seems rather narrow in comparison with that expressed in the *Code of Ethics for United States Government Service*, adopted by Congress in 1958, from which we quote the opening words:

Put loyalty to the highest moral principles and to country above loyalty to persons, party, or Government department.[10]

We submit that this higher loyalty implies a commitment on the part of government science advisors to provide their fellow citizens with the information and analyses necessary for effective participation in the political process. As a reminder of the fundamental democratic principles which are occasionally forgotten in the practical business of running the government, we offer here some simple guidelines on the limits of advising confidentiality:

1. The advisor has the obligation to bring to public attention government policies or practices that he believes may threaten the public health and welfare.
2. The advisor has the responsibility to speak out when he believes that public debate is being needlessly hampered by the misrepresentation or suppression of information.

We do not propose that our guidelines be engraved in stone. Our purpose is rather to stimulate discussion of the issues involved. Hopefully the advisors themselves and the scientific community as a whole will (perhaps with some prodding) define a new role for the scientific advisor which emphasizes his larger responsibilities.

To make the relevance of the guidelines clear, let us see how they apply to the behavior of advisors in the examples discussed in the preceding chapters.

Warning the Public

In two of our case studies, advisors took their concerns to the public. After he had been invited to testify, Richard Garwin drew Congressional attention to the degradation of the quality of life in metropolitan areas which would result from the enormous takeoff noise of the SST and to the technological setbacks which had compromised the plane's design. And, at the beginning of the public debate over the Sentinal ABM System, two governmental advisors, Hans Bethe and Richard Garwin, presented in a *Scientific American* article[11] the arguments which led them to believe that building an ABM system designed to defend the population of the United States was futile. (Defense Department clearance for publication of this article was obtained only at the authors' insistence, and not without some duress.[12])

These are the only examples in our case studies where advisors took the initiative in making their concerns public. In general, advisors remained silent—or, at most, muttered a little. The members of the Food and Nutrition Board of the National Academy of Sciences, for example, seem to have displayed a forbearance which can only be compared with that of Job while, for almost fifteen years, the Food and Drug Administration ignored their repeated expressions of concern about widespread public consumption of cyclamates.

Correcting the Record

In our case studies, advisors directly contradicted statements by administration spokesmen only when misquoted *by name*. Thus, for example, in the ABM debate, when Dr. Panofsky's name was taken in vain by Deputy Secretary of Defense Packard, and later, when Drs. Drell and Goldberger's confidential advice was misrepresented by Director of Defense Research and Engineering John Foster, a confrontation became unavoidable. In another case, Garwin, as we have already noted, tried to set the record straight—not by directly contradicting government statements, but by trying to set the actual technical basis for the decision at issue before the Congressional committees concerned.

It is not an infrequent occurrence for confidential government reports which contradict the statements of administration spokesmen to be "leaked" to the media. For example, the advisory report to the Environmental Protection Agency on the safety of 2,4,5-T entered the public domain without official approval. In this case the "leaker" was presumably concerned because the report uncritically dismissed serious concerns about possible dangers to public health. If so, his tactic was effective: as a result of criticisms of the leaked report by independent scientists, EPA administrator Ruckelshaus rejected its recommendations that 2,4,5-T be given a clean bill of health. This was an unusual case, however, in that there was a qualified group, the Committee for Environmental Information, outside government which immediately picked up and articulated the issues involved. In most cases one cannot expect a leaked report to be as influential as an advisor who himself draws the spotlight to the existence of a suppressed report and speaks to the broader implications of its conclusions. Even less useful than a leaked report is leaked advice without supporting documentation. For example, President Nixon's ad hoc SST Advisory Committee's negative views of the SST were accurately reported in the *New York Times* in March 1969,[13] but it was not until the documents themselves were released seven months later that widespread public interest was generated.

These remarks are not intended to discourage leaking to the media information that the public is legitimately entitled to have. We are simply restating our belief that scientific advisors should act more often to take the issue of suppression of information directly to the public. Excessive dependence upon Ralph Nader and the media in these matters reflects badly on the integrity of the scientific profession.

NOTES

1. Quoted in Studs Terkel, "Servants of the State: a Conversation with Daniel Ellsberg", *Harpers*, February 1972, p. 52.

2. The "socialization of the science advisor is discussed along with many other problems of the executive science advisory system by Martin L. Perl, "The Scientific Advisory System: Some Observations," *Science* 173 (1971): 1211.

3. Jason Division, Institute for Defense Analysis, "The Effects of U.S. Bombing on North Vietnam's Ability to Support Military Operations in South Vietnam: Retrospect and Prospect," August 29, 1966, reprinted in part in Neil Sheehan et al., *The Pentagon Papers* (New York: Bantam Books, 1971), pp. 502-9. A follow-up Jason study in December 1967 again concluded that the bombing of North Vietnam was militarily ineffective; see *The Pentagon Papers* (Boston: Beacon Press, 1971), Vol. 4, pp. 222-225, 231.

4. Robert S. McNamara, memorandum for President Johnson, "Actions Recommended for Vietnam," October 14, 1966, reprinted in Sheehan, *The Pentagon Papers*, pp. 542-551.

5. Sheehan, *The Pentagon Papers,* pp. 483-485. A more complete discussion of the organization of the Jason study is given by Deborah Shapley, "Jason Division: Defense Consultants Who Are Also Professors Attacked," *Science* 179 (1973): 459. Also illuminating is Eleanor Langer, "After the Pentagon Papers: Talk with Kistiakowsky, Wiesner," *Science* 174 (1971): 923-928.

6. See, e.g., Paul Dickson and John Rothschild, "The Electronic Battlefield: Wiring Down the War," *Washington Monthly*, May 1971, p. 6.

7. After a year's delay the report was forced out into public view by the Senate Foreign Relations Committee. The report was published in U.S. Atomic Energy Commission, *Underground Nuclear Testing*, AEC Report No. TID 25180 (Washington, D.C.: AEC, September 1969). The quote appears on p. 52. For a fascinating glimpse into how the advisory system was used and abused in this case see the hearing: U.S. Senate, Committee on Foreign Relations, *Underground Nuclear Weapons Testing*, 91st Congress, 1st Session, September 29, 1969.

8. "Guidelines for the Practice of Operations Research", *Operations Research: The Journal of the Operations Research Society of America* 19 (1971) p. 1123. This issue of *Operations Research* was devoted mainly to an attack on the Congressional testimony of several scientists (not members of ORSA) who opposed the Safeguard ABM system. ORSA's criticism bases a broad and unjustifiable condemnation of the ABM critics on their handling of a very narrow technical issue. For a detailed critique of the ORSA position see Paul Doty, "Can Investigations Improve Science Advice—the Case of the ABM," *Minerva* 10 (1972): 280.

9. "Guidelines for the Practice of Operations Research," p. 1134.

10. U.S. Congress, *House Concurrent Resolution 175*, 85th Congress, 2nd Session, 1958.

11. Richard L. Garwin and Hans Bethe, "Anti-Ballistic Missile Systems", *Scientific American*, March 1968, p. 21.

12. The article was partly based upon a talk by Bethe and Garwin at the annual meeting of the American Association for the Advancement of Science in December 1967. Anne Cahn in "Eggheads and Warheads: Scientists and the ABM" (Ph.D. dissertation, MIT, Department of Political Science, Science and Public Policy Program, 1971), p. 91, states that "Bethe claims he spent the last ten days before his scheduled talk on the phone, urging Defense officials to clear it [and that Director of Defense Research and Engineering] John Foster gave up a Saturday golf date on December 23 to clear the article personally." Cahn quotes Garwin as saying that he submitted his talk to John Foster only "for comment, not clearance, and received guidance on questions of classification regarding thermonuclear weapons."

13. *New York Times*, March 16, 1969. p. 1.

CHAPTER 9

Toward an Open Advisory System

It no longer suffices for me to call a group of scientists to my office and, when we have finished, to announce that based on their advice I have arrived at a certain decision. Rather it is necessary for me to lay my scientific evidence and advice on the table where it may be examined and, indeed, cross-examined by other scientists and the public alike before I make a final decision. [1]

—William Ruckelshaus,
Administrator of the
Environmental Protection Agency,
after scientific criticism of a
leaked advisory report resulted
in EPA rejection of the
report's recommendations

Congress, the public, and the scientific community have allowed the federal executive branch to establish a system of science advising committees whose activities and reports have usually been kept confidential—except when it has suited the purposes of a particular department or agency to make them public. This has made it possible for government spokesmen to create the public image that federal policies for technology follow directly from the facts and analyses provided by technical experts, even when these policies have been in reality politically motivated and technically misguided. It is intolerable that the government advising system has been so easily subverted and turned into a propaganda device for tranquilizing instead of informing public opinion. Democracy is meaningless in the absence of an informed public.

As the last chapter stressed, the integrity and effectiveness of the advising system rests in part on the willingness of individual advisors to defend it against abuse. But exhortations to advisors are not enough. Government science advisors are by and large individuals of personal integrity who try to serve their fellow citizens to the best of their ability. If they do not always succeed, that is more often the result of faulty institutions and procedures than of corrupt individuals. Governmental institutions—particularly the confidential nature of the advising system—should be reformed so as to buttress, not undermine, the personal responsibility of advisors. Fortunately, a limited but important step in this direction was taken in 1972 when, after two years of hearings, Congress finally passed the Federal Advisory Committee Act (Pub. L. 92-463). This new law has already had a considerable impact on a large proportion of federal advisory committees.

The most immediate effect has been to make these advisory committees visible. There is now a new category of announcements in the federal executive branch's official "bulletin board," the daily *Federal Register*: "Meetings." These items, of the order of ten a day, announce the meeting of various government advisory committees, at least half of which we would call scientific advisory committees. The announcements indicate the name of the committee, the purpose of the meeting, the time and place of the meeting, what parts if any of the meeting will be closed to the public and the reasons for such closure, and the required procedures for submitting written (and sometimes oral) presentations for the committee's consideration.

Inspection of these announcements often reveals that reasons cited for excluding the public are one or another of the exemptions to the Freedom of Information Act of 1967 (5 U.S.C. 552)—exemptions which also apply to the Advisory Committee Act. Sometimes meetings directly concerning the public health and welfare are closed, such as discussions of the safety and efficacy of particular drugs or the safety of particular nuclear power plants. But this does not mean that the Advisory Committee Act is useless. The fact remains that the principle of openness has been written into law, the public is informed of the meetings, and interested parties can threaten to go to court if they think that meetings are being improperly closed. Such a threat by one of Ralph Nader's lawyers was effective, for example, in getting parts of the meetings of the Atomic Energy Commission's Advisory Committee on Reactor Safeguards opened to the public.

Another useful provision of the Advisory Committee Act stipulates, subject again to the exemptions to the Freedom of Information Act, that

the records, reports, transcripts, minutes, appendixes, working papers, drafts, studies, agenda, or other documents which were made available to or prepared for or by each advisory committee shall be available for public inspection and copying at a single location in the offices of the advisory committee or the agency to which the advisory committee reports.[2]

The Advisory Committee Act has clearly effected a fundamental change in the context of government science advising. Even more important, perhaps, is the increased public skepticism regarding administration pronouncements that has resulted from the Pentagon Papers and Watergate episodes with the attendant revelations of the extent of systematic government deception of Congress and the public.

In the future, it will certainly be more difficult than it has been in the past for executive-branch agencies to misuse their advisory committees. But the subversion of the advisory system has a certain timeless quality, and we expect that, as long as governments receive advice, attempts will continue to be made to exploit the advisors and their advice for political purposes. A new law, an altered advisory structure, even a new public appreciation of democracy following a close call for the republic—none of these change the fact that technical advice will always be needed and that political advantage will always be sought by the administration. New developments do not mean that the battle for an open advisory system has been won—only that it will have to be fought on somewhat different terms.

That part of the battle waged using the provisions of the Advisory Committee Act will increasingly take place in the courts. But the new act is a very limited legal instrument. Part of its problem lies with the vagueness of some of the exemptions to the Freedom of Information Act. Because there are no provisions for punishing those who abuse these exemptions, because only exceptional pieces of information are worth the trouble and expense of the legal process, and because even such information will probably be much less valuable by the time judicial procedures are completed anyway, the arbitrary denial of information by government officials and bureaucrats is virtually risk-free.

A second weakness of the Advisory Committee Act is that it can be interpreted to apply only to advisory committees directly appointed by government officials. In particular, the advisory committees whose services are contracted for through the National Research Council (NRC) of the National Academies of Science and Engineering appear to be entirely exempt from the act's provisions. The NRC fields some 800 advisory committees, with a total membership of about 8,000 scientists (of whom only about 225 are members of the National Academies themselves). These comprise nearly one-half of the entire executive-branch science advisory establishment.[3]

In 1970 the National Academy of Sciences (NAS), following its embarrassing experience with the NAS-NRC report on the possibility of damage from SST sonic booms (see p. 54), established a Report Review Committee. The new review procedure involving this committee has prevented some obviously biased advisory reports from seeing the light of day in their original form. And the NAS does its best to see that the final reports are not suppressed for illegitimate reasons. While this self-policing is laudable, it does not diminish the importance of the openness provisions of the Advisory Committee Act. Furthermore, having different criteria of openness for different advisory committees may encourage secrecy-minded bureaucrats to "shop around."

Because of the legal complexities of the Advisory Committee and Freedom of Information acts and because of their inapplicability to research done under contact by nongovernmental concerns, the confidentiality of the advisory committee system promises to remain a problem for a long time. For this reason it is worth discussing the traditional arguments that have been used to justify advisory confidentiality, many of which were set forth as reasons for adopting the exemptions to the Freedom of Information Act.

The Arguments for a Closed Advisory System

Four principal reasons have been put forward for limiting the openness of the executive branch's system: (1) advisors must be protected from pressure or retaliation; (2) executive-branch officials must be able to have frank and open discussions with advisors during the policy-making process; (3) the confidentiality of advice and information relating to military technology which could be of use to a potential enemy must be assured; and (4) commercial trade secrets and personal privacy must be safeguarded. Each of these has some degree of merit and must therefore be weighed in individual cases against the general arguments for openness.

PROTECTING THE ADVISOR

President Nixon, when asked why the SST Advisory Committee report and the Garwin Report on the SST were being kept secret, explained in a news conference:

I have no objection to the substance of reports being made public. The problem here is that, when reports are prepared for the President, they are supposed to be held in confidence. And some of those who participate in the making of those reports have that assurance.[4]

A dozen years earlier, a similar official explanation was given by President Eisenhower to Senator Lyndon Johnson when the latter demanded the release of the Gaither report on U.S. military preparedness:

From time to time the President invites groups of specially qualified citizens to advise him on complex problems. These groups give this advice after intensive study, with the understanding that their advice will be kept confidential. Only by preserving the confidential nature of such advice is it possible to assemble such groups or for the President to avail himself of such advice.[5]

That such explanations are not always totally honest is shown, for example, by the fact that the members of the Gaither committee themselves were pressing for the release of their report in a "sanitized" version (i.e., with military secrets omitted).[6] Similarly, the report of the PSAC panel on the Safety of Underground Nuclear Testing was kept secret despite the panel's explicit recommendation that it be released.[7] The desire to avoid giving their critics ammunition is a more plausible explanation for Presidents' unwillingness to release such advisory reports.

There are certainly circumstances where advisors might fear retaliation from their employers or funding agencies if the substance of their advice became known. For example, if the chairman (as of 1973) of the NAS-NRC committee advising the Defense Department on cereal and general products, who happens to be employed by the ITT-owned Continental Baking Company, were ever to advise the government that his company's products are in some respect inferior to a competing brand and this information became known to ITT, it is possible that his future prospects at Continental would be somewhat diminished. Problems like this obviously should be minimized by choosing advisory committee members so as to minimize conflicts of interest.

Another, more legitimate concern that might bother an advisor is that he will receive unwelcome attention if it becomes widely known that he is advising on some currently controversial issue. For example, some Columbia University physicists who were members of the elite Jason group of summer consultants to the Defense Department found themselves being harassed by threatening telephone calls and hate mail because of the group's work on weapons technology for the war in Indochina.[8]

An earlier and much more serious example is the case of J. Robert Oppenheimer, the physicist who led the atomic bomb project during World War II. In 1954 Oppenheimer was called before an Atomic Energy Commission hearing board, stripped of his security clearance, and politically disgraced—mostly on the basis of charges twice previously investigated and dismissed as relatively unimportant. The belief is widespread among scientists that Oppenheimer was persecuted because he became too highly visible as a government advisor and because elements in the military who disagreed strongly with his advice on strategic weapons wanted to destroy his influence.[9]

These have been unhappy episodes, and we would be the last to wish to see them repeated. However, an overly protective attitude toward advisors would only engender more abuses of the sort documented in our case studies. Government officials with important public responsibilities are expected to be answerable to the people for the way they carry out these responsibilities. If science advisors are unwilling to take public responsibility for their participation in government decision making, the seriousness of their dedication to the public interest comes into question. In the last analysis, the support of the scientific community and the confidence of the public in the integrity of the policy-making process seem to be the best and most appropriate guarantees of the political independence of the science advisor.

FULL AND FRANK DISCUSSIONS

If the public were given access to every discussion within the executive branch, the result would be quite disruptive. Certainly the ability of officials to participate in "full and frank discussions" during the governmental policy-making process would be inhibited. On the other hand, if executive deliberations were entirely insulated from the press and public, the only external voices heard in these deliberations would be those of large Presidential-campaign contributors and other well-connected parties. Clearly some middle ground must be sought between complete openness and complete secrecy. Traditionally, Presidents and other executive-branch officials have leaned in the direction of secrecy.

The most potent device the President can use to resist requests from Congress for executive-branch documents is the invocation of "executive privilege." Executive privilege is legitimately supposed to protect delicate matters such as ongoing international negotiations and the President's personal consultations. The President's immediate full-time staff is unually also considered to be shielded by the umbrella of executive privilege—at least to the extent that they act as the President's personal agents and advisors. But the wholesale extension of this doctrine to include large numbers of documents prepared by groups of part-time advisors is unjustifiable. In cases like that of the Garwin Report on the SST, needless confidentially has denied Congress and the public timely access to the only comprehensive and authoritative studies in existence.

Early in his Presidency, Dwight Eisenhower issued a directive to his Secretary of Defense in which he gave his interpretation of the justification for and extent of executive privilege:

> Because it is essential to efficient and effective administration that employees of the Executive Branch be in a position to be completely candid in advising with each other on official matters, and because it is not in the public interest that any of their conversations or communications, or any documents or reproductions, concerning such advice be disclosed, you will instruct employees of your Department. . . not to testify to any such conversations or communications or to produce any such documents or reproductions. . . .
>
> I direct this action so as to maintain the proper separation of powers between the Executive and Legislative Branches of the government in accordance with my responsibilities and duties under the Constitution.[10]

This statement was construed by many executive agencies as justifying almost any refusal of information that may be requested by Congress.

President Eisenhower issued his directive in a period when Senator Joseph McCarthy's (R.-Wisc.) investigations had induced a state of near-paranoia in the executive branch. Since that time, Presidents Kennedy, Johnson, and Nixon have each expressly repudiated the applicability of executive privilege to the whole executive branch, affirming that this power may be invoked only by the President himself.[11] But these fine promises have not always been observed—most notably during the Watergate affair, when Attorney General Richard Kleindienst testified at one point that the President could, if he wanted to, apply executive privilege to the entire executive branch, and that if the Congress did not approve of this policy, its only recourse was to impeach the President.

The legal status of executive privilege remains obscure because the issue seldom comes to court. Except for litigation, the only limitation on what the President can get away with in withholding information is Congressional and public outrage. Consequently, the invocation of executive privilege has long been a Congressional irritant. In 1960 the House Government Information Subcommittee commented:

The great bulk of requested documents are eventually released, but the questionable doctrine of Executive privilege results in unwarranted delay. Because of the timing of legislation and the shortness of the sessions of the Congress, delay is frequently tantamount to complete obstruction, preventing the timely exposure and correction of executive branch errors.[12]

As far as Congress is concerned, executive privilege at most extends to the President's personal consultations on matters of state. The only restrictions on full disclosure of the deliberations and memoranda of lower-ranking executive-branch officials are the explicit exemptions written into the Freedom of Information Act. In cases not covered by standard exemptions like military security, the last refuge of a reticent bureaucrat is "exemption 5," which exempts

interagency or intra-agency memorandums or letters which would not be available by law to a party other than an agency in litigation with the agency.

According to the interpretation of the federal Office of Management and Budget, this exemption applies also to the verbal discussions of advisory committees which would be covered were they written down, if, in addition, the agency head determines that

it is essential to close such meeting (or portion) to protect the free exchange of internal views and to avoid undue interference with agency or committee operation.[13]

Unfortunately, the exact legal meaning of these provisions is not entirely clear. The primary function of advisory committees should be to discuss the factual and analytical bases for a decision—and it would seem that these should ordinarily be made as freely available to the concerned citizen as to the government official. Certainly the wholesale concealment of advisory reports, such as those on the safety of underground nuclear tests in the Aleutian island of Amchitka or the Garwin Report on the supersonic transport, is not required by any general considerations of good government. Judicial opinions regarding "exemption 5" in two representative cases seem to support this view.[14] In *Mink et al.* v. *EPA* (Amchitka), the district court held that

while the exemption protects the decisional processes of the President, or other policy-making executive officials, it does not prevent the disclosure of factual information unless it is inextricably intertwined with policy making processes.[15]

In *Soucie* v. *David*, concerning the release of the Garwin Report, the appeals court gave a similar interpretation of the law:

Factual information may be protected only if it is inextricably intertwined with policy-making process. . . . [The] courts must beware of the inevitable temptation of a government litigant to give [this exemption] an expansive interpretation in relation to the particular records at issue.[16]

In view of these opinious, the legitimate applicability of "exemption 5" to the deliberations of advisory committees would appear to be rather small.

Accordingly, it might be appropriate for Congress to make this exemption inapplicable to advisory committees by law.

SECURITY CLASSIFICATION

"Exemption 1" under the Freedom of Information Act applies to matters specifically required by Executive order to be kept secret in the interest of the national defense or foreign policy.[17]

Although modest in length, this exemption leaves inviolate the whole bureaucratic nightmare which goes by the name of "security classification."

The present system of security classification has few defenders. In a 1970 report to the Secretary of Defense, the Defense Department's own Defense Science Board Task Force on Secrecy estimated that "the volume of scientific and technical information that is classified could profitably be decreased by perhaps as much as 90 percent through limiting the amount of information classified and the duration of its classification."[18] Even the National Security Council official responsible for drawing up revised security classification procedures in the wake of the Pentagon Papers incident admitted that there was a problem: "We are trying to reverse 20 years of practice under which there were abuses in overclassification."[19]

Unfortunately, the Nixon administration's revision of the classification procedure is not very convincing. Its major new provision is automatic declassification of documents after a certain number of years; but even documents that are merely classified "confidential" (the lowest security classification) must wait six years before automatic declassification. Furthermore, this "automatic" declassification is subject to bureaucratic review, the final authority in case of disputes over classification being the Interagency Classification Review Committee—consisting of members of the agencies which classify documents. It is no wonder that Representative William S. Moorhead (D.-Pa.) criticized Nixon's executive order establishing the new system as "a document written by classifiers for classifiers."[20]

One merit of the executive order, however, is that it includes a capsule description of the abuses of security classification which should be prevented:

In no case shall information be classified in order to conceal inefficiency or administrative error, to prevent embarrassment to a person or Department, to restrain competition or independent initiative, or to prevent for any other reason the release of information which does not require protection in the interest of national security.[21]

One obvious measure to prevent such abuses would be for Congress to set up an independent review board with the power to hear and rule on classification matters. Its services should be available to help members of Congress, the press, and the public locate and obtain information to which they are legitimately entitled. Hopefully the time will come when a citizen has reasonably prompt recourse when he is told, for example, that the findings of a survey on the incidence of birth defects in Vietnam has been "classified." (See page 158.)

TRADE SECRETS AND CONFIDENTIAL MATERIAL

A final justification frequently used to defend closure of science advisory committee meetings is "exemption 4" of the Freedom of Information Act, regarding

trade secrets and commercial or financial information obtained from a person and [matters which are] privileged and confidential.[22]

Here, as with the other exemptions, respect for the rights of individuals and businesses must be balanced with the social concern for freedom of information. Two examples will give an idea of the types of cases in which this issue arises:

1. Committees which advise the National Institutes of Health close their meetings during discussions of the abilities of particular scientists and the merits of their research proposals. Here, a proper respect for the privacy of the individual researcher must be balanced against society's concern that the taxpayers' money be well spent. It is difficult to decide this balance on general principles. Most scientists believe that the peer review system is currently working in the public interest.[23]

2. The Food and Drug Administration closes advisory committee meetings in which the safety and effectiveness of particular drugs are discussed—arguing that among the relevant information are trade secrets. The Atomic Energy Commission uses the same argument to justify the closing of those parts of meetings of its Advisory Committee on Reactor Safeguards which discuss the effectiveness of key reactor safety systems on particular reactors.[24] In both cases it would seem that the public interest in seeing that these safety issues are properly handled is so overwhelmingly great that secrecy should not be tolerated. If necessary, the Freedom of Information Act should be amended to make this clear.

Problems of Bias in Advisory Committees

The Advisory Committee Act gives no guidance on issues relating to the membership of advisory committees other than to specify that the names and occupations of each committee's members be published in an annual report to Congress. Presumably the architects of the act felt that the provisions of openness it contains would expose problems of bias and conflict of interest to public view and thereby tend to bring about corrective action. And many executive agencies, and also the National Academy of Sciences, have recently established procedures for eliminating obvious bias and conflict of interest in their advisory committees.

These problems are persistent and subtle, however.

For one thing, any committee made up solely of experts in a particular subject is likely to be biased from the outset. People used to working and thinking in a certain discipline, and who thus tend to see issues in the context of that discipline, inevitably base their advice on a certain set of implicit technological, social, and political assumptions. Hugh Folk has described the problem:

It is inevitable that experienced experts will usually be drawn from the interests involved in a problem. In many instances the experts will have created the problem. The A.S.E.B. [Aeronautics and Space Engineering Board of the National Academy of Engineering] appears to be incapable of entertaining an idea injurious to air transport. Just as automotive executives and engineers could not generate any interest in auto safety, so these men cannot generate any interest in quiet. They perceive the problem in terms of "tolerable noise."[25]

Obviously such a bias should 'be compensated by including members with qualifications other than expertise in the "offending" technology.

Another way in which bias is introduced into an advisory committee is through the exclusion of individuals who have taken strong public stands on the matters at issue. At first sight such a procedure might seem neutral and in the interests of an effective committee. Decision makers want advice, not unresolved arguments (it is explained), and persons with strongly held views will not easily be persuaded to join a consensus.

Unfortunately, the exclusion of such individuals automatically results in a bias toward the status quo. In public controversies about technical issues, scientists who disagree with established policies have to raise their voices merely to be heard, while scientists who support existing policies encounter little such resistance—if they feel the need at all to add their voices to those of official government spokesmen. Consequently, an advisory committee made up of "moderates" often lacks a spokesman for the very criticisms that may have prompted the convening of the committee in the first place. Characterizing scientific critics of established policies as "contentious," "unreasonable," "uncompromising," or "disruptive" is one of the most unfortunate by-products of public controversies over technology. In interviewing a substantial number of these "controversial" scientists in researching this book, we have found their most distinctive characteristic to be not contentiousness but rather the self-confidence and lack of awe for authority which are obvious prerequisites for individuals who are going to stand up and effectively articulate nonestablishment positions in the public arena. It is a considerable loss to society for such individuals to be systematically excluded from advisory committees after they have taken a public stand.

In 1972 a prestigious National Academy of Sciences committee (whose members included two former presidents of the NAS) was commissioned by the Advanced Research Projects Agency of the Department of Defense to look into the problem of identifying and recruiting young advisors. The committee's report described the standard procedure—the "telephone method" or "buddy system"—as follows:

Staff members, members of an executive committee, or others assigned to this activity in the responsible organization call professional colleagues or write to them describing the committee's task and soliciting suggestions of candidates.

Those usually asked to make nominations are people with established reputations in the field, who often have served as members or chairmen of committees. Their judgement is respected by the sponsoring organization. Cross-checking and collection of further information about nominees follow. The list of names of nominees is screened repeatedly as the requirements become better established, until a group of persons who meet the dominant criteria has been selected.[26]

Although the committee reported that this selection procedure basically works well, it did acknowledge that "it tends to call upon 'the same old faces' repeatedly." Methods for improving the search procedure were suggested but with so little conviction that the NAS itself has essentially ignored them.

One method for broadening the membership of important advisory committees to include scientists who could make valuable contributions but who might not come to the attention of the ordinary "buddy system" is to publish in relevant magazines a notice of the charge to the committee and at the same time solicit suggestions as to how the committee might best go about carrying out that charge. This would help to identify people interested in and actively thinking about the question at hand; and obviously some procedural suggestions might be very useful. A magazine like *Science* might appropriately carry such notices, and the news magazines of professional societies could also publish those notices which might be of special interest to their members. Another method might be for the professional societies themselves to circulate questionnaires among their members asking whether they would like to advise or do other work *pro bono publico,* and if so, in what areas. On the basis of the replies, a committee might be established to nominate members for particular advisory committees. Beyond this, the professional societies should encourage discussion—at their meetings and in their publications—of the responsibilities of advisors—especially in light of the provisions of the Advisory Committee Act.

Conclusions

We have argued here the importance of further drawing aside the curtain of confidentiality behind which executive-branch advising and decision making have too long been hidden. Besides making important information available to those who need it both inside and outside government, free access to advisory reports and proceedings will almost inevitably improve the quality of the advice—because data and judgments would be subjected to the scrutiny of free scientific debate; because the various practices by which officials attempt to influence advice, from "packing" of committees to intimidation of advisors, would become less practicable; and also because creative proposals and thoughtful judgments would redound to the credit of their authors.

Scientists are always rightly suspicious of any scientific claims or conclusions which are presented without adequate supporting evidence. There is no reason why this fundamental tenet of the scientific method should not apply equally to the technical advice and analyses on which public policy is based.

NOTES

1. Quoted in *Science* 174 (1971): 43.

2. Federal Advisory Committee Act (Pub. L. 92-463; 86 Stat. 770), Sec. 10(b).

3. *The Science Committee, a Report by the Committee on the Utilization of Young Scientists and Engineers in Advisory Services to Government* (Washington, D.C.: National Academy of Sciences, 1972). See especially Appendix, Table D-1. See also *NAS-NAE-NRC Organization and Members* (Washington, D.C.: National Academy of Sciences, annual), and John Walsh, *Science* 172 (1971): 242 and 343.

4. *Public Papers of the Presidents of the United States: Richard M. Nixon, 1970* (Washington, D.C.: Government Printing Office, 1971), pp. 1104-5. Quoted in the *New York Times,* December 11, 1970, p. 32.

5. Quoted in *ibid.,* January 23, 1958, p. 10.

6. See Morton H. Halperin, "The Gaither Committee and the Policy Process," in *The Presidential Advisory System,* ed. Thomas E. Cronin and Sanford D. Greenberg (New York: Harper & Row, 1969), p. 199. The Gaither committee—composed of prominent industrialists and scientists and chaired by H. Rowan Gaither, Jr., chairman of the boards of the Ford Foundation and the Rand Corporation—was assembled in 1957 to advise the National Security Council on the need for fallout shelters and on other defense problems. Their report warned of growing Soviet military strength and recommended crash programs to build fallout shelters and to decrease the vulnerability of American strategic weapons. The Gaither report was finally released in 1973 as a result of an appeal by the *New York Times* to the newly established Interagency Classification Review Committee (see p. 119). See the *New York Times,* January 20, 1973, p. 1.

7. This report—known as the "Pitzer report" after Kenneth Pitzer, who chaired the PSAC panel—was released only three days before the "Milrow" nuclear weapons test at Amchitka Island, after strong pressure had been brought to bear by the Senate Foreign Relations Committee. See U.S. Congress, Senate, Committee on Foreign Relations, *Underground Nuclear Weapons Testing,* 91st Cong., 1st Sess, September 29, 1969, esp. pp. 31ff. The Pitzer report is included in U.S. Atomic Energy Commission, *Underground Nuclear Testing,* Report No. TID 25180, September 1969, pp. 51 ff.

8. In addition, during the summer of 1972, demonstrations by students in Europe prevented two prominent physicists from giving scheduled lectures and even resulted in the premature closing of the summer school at which one of them was to be a lecturer. See Deborah Shapley, *Science* 179 (1973): 459.

9. The dramatic transcript of the Oppenheimer security hearing along with the texts of the principal associated documents and letters are reprinted in U.S., Atomic Energy Commission, *In the Matter of J. Robert Oppenheimer* (Cambridge, Mass.: MIT Press, 1970).

10. Letter from President Eisenhower to the Secretary of Defense, May 17, 1954. Quoted in Charles R. Dechert, "Availability of Information for Congressional Operations," in *Congress: The First Branch of Government,* Alfred de Grazia, ed. (New York: Anchor, 1967), p. 182.

11. Regarding the statements of Presidents Kennedy and Johnson, see *ibid.,* p. 183. Such a pledge by President Nixon was noted by Representative Reuss in the *Congressional Record* 115 (1969): 34743.

12. U.S. Congress, House, Committee on Government Operations, *Executive Branch Practices in Withholding Information from Congressional Committees,* House Report No. 2207, 86th Cong., 2nd Sess., August 30, 1960, p. 4.

13. Draft Joint OMB/Department of Justice Memorandum on Implementation of the Federal Advisory Committee Act, Section 10 a(3)(c)(iii)(c). This memorandum is included as Appendix D to *Federal Advisory Committees: First Annual Report of the President* (Washington, D.C.: Office of Management and Budget, March 1973). The latter document also includes a list of current government advisory committees. More detailed information, including membership, meetings, and publications, is included in U.S. Congress, Senate, Committee on Governmental Operations, *Federal Advisory Committees, First Annual Report of the President to Congress, Including Data on Individual Committees,* 93rd Cong., 1st Sess., May 2, 1973, 4 vols, 5703 pp. See also Linda E. Sullivan and Anthony T. Kruzas, eds., *Encyclopedia of Governmental Advisory Organizations* (Detroit: Gale Research Co., beginning July 1973).

14. A summary of judicial decisions relating to the Freedom of Information Act may be found in U.S. Congress, House, Committee on Government Operations, *U.S. Government Information Policies and Practices—Administration and Operation of the Freedom of Information Act, Part 4,* 92nd Cong., 2nd Sess., March 1972, pp. 1347-1367. See also Nicholas Wade, *Science* 175 (1972): 498. The Freedom of Information Center, Box 858, University of Missouri, Columbia, Mo. 65201, responds to requests for information on the Freedom of Information Act. See *Saturday Review,* March 31, 1971, p. 92.

15. Quoted in *U.S. Government Information Policies and Practices* (above reference), p. 1363.

16. Quoted in ibid., p. 1365.

17. 5 *U.S.C.* 552 (b)(1).

18. U.S. Department of Defense, Office of the Director of Defense Research and Engineering, *Report of the Defense Science Board Task Force on Secrecy,* July 1, 1970, p. 2. This task force was chaired by Frederick Seitz; other members were Alexander H. Flax, William G. McMillan, William B. McLean, Marshall Rosenbluth, Jack P. Ruina, Robert L. Sproull, Gerald F. Tape, and Edward Teller—a substantial fraction of the Defense Department's high-level science advisory establishment, including the current Defense Science Board chairman (Tape) as well as his predecessor (Seitz). Their recommendations were received favorably by Deputy Secretary of Defense Packard and have been partially implemented, according to letters of March 1 and October 26, 1972, from DSB Executive Secretary Leon Green, Jr., to one of the present authors (J.P.).

19. David R. Young, quoted in *National Journal,* April 5, 1972, p. 657.

20. Quoted in *The New Republic,* July 22, 1972, p. 8.

21. Executive Order No. 11652, Sec. 4, March 10, 1972.

22. 5 *U.S.C.* 552 (b)(4).

23. For scientists' responses to recent Nixon administration attacks on the peer review system of the National Institutes of Health, see Barbara J. Culliton, *Science* 180 (1973): 843 and 1035.

24. For typical examples, see U.S. *Federal Register,* Vol. 38, March 1, 1973, pp. 5496ff (FDA committees); *ibid.,* September 4, 1973, pp. 23814ff, and September 11, 1973, pp. 24937ff (AEC Advisory Committee on Reactor Safeguards).

25. Hugh Folk, "The Role of Technology Assessment in Public Policy," in *Technology and Man's Future,* ed. Albert H. Teich (New York: St. Martin's Press, 1972), p. 250.

26. *The Science Committee* (ref. 3, above), Appendix E, p. 62.

PART IV

The People's Science Advisors—Can Outsiders Be Effective?

You've convinced me. Now go out and bring pressure on me.

—President Franklin Roosevelt
as quoted by Saul Alinsky
in the Prologue of
Rules for Radicals (1971)

INTRODUCTION

THE executive branch of our government has not been giving citizens the technical information they need. Scientists must therefore make their expertise directly available to the public and Congress.

The idea that the public as well as government and industry should have scientific advisors is an old one—as is the idea that the public interest should have lawyers to defend it. It was not until the 1960s, however, that a renewed public understanding of the insensitivity of government and industrial bureaucracies led to a substantial commitment in the legal profession to public interest law. It appears that the scientific community may now have reached a similar point. The growing public awareness of the dangerous consequences of leaving the exploitation of technology under the effective control of special industrial and government interests has led to a readiness within the scientific community to undertake a serious commitment to what we have termed "public interest science."

There is an important difference between the practice of public interest law and public interest science, however. In a legal dispute, once both parties have obtained lawyers they can hope to receive a fair and equal hearing in front of a trained judge who gives their arguments his undivided attention. In a public debate over an application of technology, on the other hand, tremendous inequalities exist. The contending sides must speak to a distracted public through news media to which administration officials have comparatively easy and routine access. Moreover, an executive-branch official speaks with the authority of his office, while an independent scientist is usually an unknown quantity to the public.

Thus, it is important to determine whether the public interest activities of independent scientists can in fact activate political and legal restraints on irresponsible executive-branch actions. In this section of the book we present six case studies of instances where "outsiders" have had at least partial success. In none of these cases did the public interest scientists succeed in effecting as great a change in the policies at issue as they had hoped. But in each case, public exposure of the issues led to remedial action which had been impossible to obtain by those working within the executive branch. At least as important is the fact that many of the controversies stimulated by such exposure helped initiate a political process which often had quite far-reaching impact on the approach of society and government to the technologies involved.

CHAPTER 10

The Battle Over
Persistent Pesticides:
From Rachel Carson
to the Environmental
Defense Fund

> *"The worst residue problem we have to face today is the residue of public opinion left by Rachel Carson's 'Silent Spring'."*
>
> —Representative Jamie Whitten (D.-Miss.), chairman of the House Appropriations Subcommittee on Agriculture and a pesticide enthusiast, during the March 1968 Department of Agriculture appropriations hearings.[1]

Rachel Carson's *Silent Spring*, published in 1962, dramatically portrayed the ecological damage and potential long-term human health hazards that were resulting from indiscriminate use of persistent pesticides like DDT, which linger on in the environment for years after they are used. In Chapter 3 we saw how, as a result of the furor over Miss Carson's book, the President's Science Advisory Committee was assigned to reexamine the evidence. This report concluded that there was an urgent need for studies of pesticide hazards, and that the evidence was already sufficiently strong on persistent pesticides that most uses of such pesticides should be curtailed immediately.

Many of the research recommendations of the PSAC panel on the use of pesticides were implemented and have resulted in a much better—although still rather limited—understanding of the effects of pesticides on man and his environment. The report had little impact on government regulation of pesticides, however.

Symbolic of the lack of government *action* (as distinct from sponsorship of studies) was the nonimplementation of the PSAC panel's recommendation (renewed in another PSAC report two years later[2]) to restrict the use of DDT and other persistent insecticides to the control of disease vectors. (The use of persistent pesticides for this purpose in the United States was vanishingly small.) The banning of DDT became the focus of a continuing battle in which the chemical companies, the Agriculture Department, and the political representatives of agriculture on the one side confronted a loose grouping of public interest groups, ranging from the "bird watcher" Audubon Society to the scientist-dominated Environmental Defense Fund, on the other. A steady stream of environmental disasters—ranging from the almost annual appearance of millions of pesticide-killed fish floating on the Mississippi[3] to the identification of new bird species whose populations were declining as a result of the use of DDT[4]—continued to mobilize public opinion behind the "anti-DDT" groups.

The supporters of persistent pesticides continued to have the dominant influence within the government, however. This may be seen, for example, in the Congressional relations of the opposing sides. While the anti-DDT forces had as active allies a few Congressmen such as Senator Abraham Ribicoff (D-Conn.) and Representative Fountain, the committees which these Congressmen chaired had no power over federal pesticide policies. They could only provide national forums for the critics of federal pesticide policies.

The pro-pesticide forces, on the other hand, could claim the support of the Congressmen from cotton-belt states who were the chairmen of the key House and Senate Appropriations and Agriculture Committees and subcommittees. Perhaps the most dedicated supporter of pesticides among these was Representative Jamie Whitten (D.) of Mississippi, chairman of the House Appropriations Subcommittee on Agriculture and author of the book *That We May Live* (1966). (This book, largely researched by the Agriculture Department,[5] was the answer of the pro-pesticide forces to *Silent Spring*. Its publication was subsidized by three pesticide manufacturers.[6]) Whitten, by virtue of his strategic position and the leverage given by Congress to committee chairmen, has been able to cut or increase items in the Agriculture Department's budget almost by fiat. In fact, his power to bring the Agriculture Department around to his way of thinking has caused some to dub him "the real Secretary of Agriculture."[7]

The Environmental Defense Fund

While the anti-DDT forces were being held at bay in Washington, some groups decided to confront local and state governments with the issue. Independent scientists first became seriously involved as a result of a court suit filed in the spring of 1966 against the Suffolk County (Long Island) Mosquito Control Commission by lawyer Victor Yannacone on behalf of his wife. The commission's DDT spraying program, it seems, had resulted in a fish kill in a local lake, and Mrs. Yannacone was seeking a court injunction banning the commission from further use of DDT.

To document the case against DDT, Yannacone recruited scientists from the local branch of the State University of New York at Stony Brook and from the Atomic Energy Commission's Brookhaven National Laboratory. These scientists presented the mounting evidence of both short-term and long-term devastation of many forms of wildlife resulting from massive DDT spraying programs. The National Audubon Society provided financial support.

The suit was ultimately decided on legal rather than technical grounds. The courts decided that the setting of regulations on pesticide use was a legislative rather than a judicial responsibility, and the Yannacones lost. The case had received such extensive press coverage, however, that both the public and the Suffolk County government had made up their minds about DDT. The County Board of Supervisors went on record as opposing the use of the pesticide, and the Mosquito Control Commission then announced that it would stop using DDT.[8] (Winning by force of public opinion what has been lost in the courts is a phenomenon which we will encounter frequently in our case studies of the activities of public interest groups.)

Following the Suffolk County case, Yannacone and his scientist allies decided to continue their legal battle against the excessive use of persistent pesticides. They set up the Environmental Defense Fund (EDF) and took the pesticides to court in Michigan. Frank Graham tells in *Since Silent Spring* how they fared:

In 1967 [the EDF] filed suit in western Michigan to restrain nine municipalities from using DDT for Dutch elm disease control. Again, EDF lost in court but attained its objective. The Cooperative Extension Service of Michigan State University withdrew its statewide recommendation of DDT for this program and recommended instead sanitation methods coupled with the supplementary use of methoxychlor [a nonpersistent pesticide].

Encouraged by this success, EDF expanded its court action to include another 47 Michigan municipalities. By 1968, 50 of the 56 municipalities planning to use DDT had consented to the court orders which compelled them to use alternate methods of control. At the same time, an EDF suit temporarily averted a planned application of three tons of dieldrin [a persistent pesticide related to DDT] over Michigan's Berrien County for the control of Japanese beetles. Charles Wurster [an assistant professor of biological sciences at the State University of New York at Stony Brook, and head of the EDF's scientific advisory committee] testified that from 10 to 80 birds and mammals, up to the size of cats and even sheep, would be killed for every beetle killed in this program. Though the Michigan and United States Departments of Agriculture eventually went ahead and sprayed 3,000 acres with dieldrin for what they admitted amounted to only about a single beetle per acre, EDF moved the country closer to a sane pesticide policy. The widespread publicity given both court actions publicized the strong scientific case against the persistent pesticides and publicly discredited the spray programs.[9]

In its suit to stop Michigan's dieldrin spraying program, the EDF was joined by the state Department of Conservation. "This showdown has been coming for a long time," said one Conservation Department official. Referring to the state Agriculture Department he said, "They call us 'bird lovers.' "[10]

In late 1968 the focus of EDF legal activities moved to Wisconsin. In October, the Citizen's Natural Resources Association of Wisconsin (a group of scientists and laymen) and the Wisconsin Izaak Walton League filed petitions with Wisconsin's Department of Natural Resources requesting a ruling as to whether DDT should be defined as a pollutant according to Wisconsin state law. (According to the Wisconsin statute, a pollutant is a substance "contaminating or rendering unclean or impure the waters of the state, or making the same injurious to public health, harmful for commercial or recreational use, or deleterious to fish, bird, or plant life."[11]) The Wisconsin Department of Natural Resources responded to the petitions it had received by setting up hearings to determine whether DDT should be considered a pollutant. EDF's lawyer, Victor Yannacone, and the head of EDF's scientific advisory committee, Charles Wurster, organized the presentation of the case against DDT (including local expert witnesses). The Task Force for DDT, organized by six of the chemical companies manufacturing DDT, presented the defense. The hearings, which lasted almost six months and produced more than 2,500 pages of testimony, constituted something of a national forum for the debate over DDT.

A year after completion of the testimony, on May 21, 1970, Hearing Examiner Maurice Van Susteren issued his finding:

DDT, including one or more of its metabolites in any concentration or in combination with other chemicals at any level within any tolerances, or in any amounts, is harmful to humans and found to be of public health significance. No concentrations, levels, tolerances, or amounts can be established. Chemical properties and characteristics of DDT enable it to be stored or accumulated in the human body and in each trophic level of various food chains, particularly the aquatic, which provides food for human consumption. Its ingestion and dosage therefore cannot be controlled and consequently its storage is uncontrolled. Minute amounts of the chemical, while not producing observable clinical effects, do have biochemical, pharmacological, and neurophysiological effects of public health significance. . . . Feeding tests, laboratory experiments and environmental studies establish that DDT or one or more of its analogs is harmful to raptors [birds of prey such as eagles and falcons] and waterfowl by interfering with their reproductive process and in other birds by having a direct neurophysiological effect.

Feeding tests or experiments and environmental studies establish that DDT at chronic low levels is harmful to fish by reducing their resistance to stress.

DDT and its analogs are therefore environmental pollutants within the definitions of Sections 144.01 (11) and 144.30 (9), Wisconsin Statutes.[12]

By the time Van Susteren announced this conclusion, however, events had forced action in both Wisconsin and the neighboring state of Michigan: First, in 1968 700,000 young coho salmon which were to be used to stock Lake Michigan died in Michigan state government hatcheries—apparently of DDT poisoning.[13] Then in March and April of 1969, 28,000 pounds of frozen Lake Michigan coho salmon were seized in Michigan by the Food and Drug Administration (FDA) because their fat was found to contain levels of DDT greater than legally allowed for meat (standards had not previously been set for fish).[14] Suddenly Michigan saw DDT as a threat to the $100 million spent annually in the state by visiting sport fishermen.[15]

Responding to these shocks, in mid-April the Michigan Agricultural Commission voted to ban sales of DDT in the state.[16] In July, the Wisconsin State Assembly, following suit, decided not to await Van Susteren's finding on DDT and voted 90-0 to ban DDT usage in the state in all but emergency situations.[17]

The Agriculture Department in Trouble Over Pesticides

By late 1969 the triumphs of the anti-DDT forces in the Midwest had given them enough momentum to carry their campaign back to Washington. At the same time, other events conspired to weaken the credibility of U.S. Department of Agriculture (USDA) pesticide regulation.

In September 1968 the General Accounting Office (GAO), Congress's watchdog agency, had issued a devastating report on USDA enforcement of pesticide regulations. The GAO found the enforcement to be virtually nonexistent. In 1966, for example, out of 2,751 samples of pesticides which were tested and reviewed by the Agricultural Research Service (ARS), the pesticide regulatory arm of the USDA, 750 were found to violate the law—including 562 major violations. The ARS took action in only 106 of these cases to confiscate illegal shipments, however, and even in these cases it did not bother to track down all shipments. The GAO learned that, even though "repeated major violations of the law were cited by the agency, . . . shippers did not take satisfactory action to correct violations or ignored ARS notifications that prosecution was being contemplated."[18] Moreover, the ARS had not referred a single case to the Justice Department for prosecution in thirteen years! Dr. R. J. Anderson, acting administrator of the ARS, when asked to comment on the obvious contempt with which the industry was treating the pesticide laws, said: "We believe that cooperative action by a manufacturer in recalling defective or hazardous products is the most efficient and effective means for removing such products from channels of trade."[19]

In February 1969, as a result of a second investigation, the GAO reported that the USDA was apparently not honoring the spirit of agreements which had been made in 1964 in the revamping of pesticide-registration procedures which followed the recommendations of the 1963 PSAC report. According to this agreement, the USDA was required to seek advice from the Public Health Service, an agency of the Department of Health, Education, and Welfare, on possible health hazards before registering a pesticide for use.[20] The GAO presented a case history in which this advice had been ignored. According to the summary:

We found the Agricultural Research Service registered lindane pellets for use in vaporizing devices on a continuous basis in certain commercial and industrial establishments—such as restaurants and other food handling establishments—even though there had been long-term [almost sixteen years] opposition to this practice by the Public Health Service and Food and Drug Administration, Department of Health, Education, and Welfare, as well as other Federal, State, and private organizations.[21]

In May and June of 1969, Representative Fountain held hearings of his Subcommittee on Intergovernmental Relations of the House Government Operations Committee to examine the manner in which the USDA was discharging its responsibilities for pesticide regulation. The Fountain subcommittee found that the lindane case was fairly typical of the USDA's responsiveness to advice about pesticide safety. In 1969 alone, at least 185 pesticides had been registered over PHS objections.[22] In the same vein, a Ralph Nader group later found that, of 5,052 recommendations made in 1969 by the Food and Drug Administration to the USDA for labeling changes which would encourage safer use of various pesticides, *none* were accepted.[23]

The Government Moves Against DDT (or So It Appears)

At the same time that the credibility of the USDA as a protector of the public health was being destroyed, new information indicating that DDT might be a serious human health hazard became public. In May 1969 HEW's National Cancer Institute released a report on an experiment in which it was found that DDT causes cancer in mice. During the following months decisions were made in two more states not to wait any longer for federal action against DDT: in June the New York State Pesticide Control Board asked the state legislature to curb the use of DDT in the state,[24] and in July the California State Senate voted to ban DDT in that state.[25] Arizona had already in January banned the use of DDT for a year.[26] In the light of these state decisions, the USDA's position on DDT was becoming increasingly untenable.

In July 1969, the USDA made its first moves to decrease the use of DDT in its own programs. It tied the announcement to the release of a National Academy of Sciences' National Research Council (NAS-NRC) advisory report. The report echoed the 1963 PSAC report on pesticides in recommending that "more effective steps be taken to reduce the unneeded and inadvertent release of persistent pesticides into the environment."[27] Backed up by this recommendation, the USDA announced that it was temporarily suspending its programs for spraying persistent pesticides at airports and in the national forests. This announcement was followed by another in August, in which the USDA made public its intention to stop using persistent pesticides in two federal-state insect-control programs and to drastically reduce their use in another.[28] The USDA was still unwilling to make the politically dangerous decision to limit the use of DDT in agriculture, however. Nearly two-thirds of the remaining DDT use in the United States was on cotton—and, as we have noted, the cotton growers were well-represented in Congress by Representative Whitten and the other Southern Congressmen. At least one powerful Congressman, Senator James Eastland (D.) of Mississippi, is a major cotton grower himself. (Actually, according to the 1972 *Almanac of American Politics*, Senator Eastland was receiving about $160,000 annually for *not* growing cotton.[29])

In October 1969, the Environmental Defense Fund and four other conservation groups with which it had become allied (the Sierra Club, the National Audubon Society, the West Michigan Environmental Action Council, and the Izaak Walton League of America) petitioned Secretary of Agriculture Clifford Hardin to suspend all registered uses of DDT.[30] Their petition went unanswered.

The tide of national and international opinion continued to turn against DDT and the USDA position continued to erode, however. In October, California took action to cut the use of DDT in the state by about one-half,[31] and in early November the Canadian government—which also found itself under pressure as a result of the actions of several provincial governments which had limited the use of DDT—announced restrictions which would cut the use of DDT in that country by an estimated 90 percent.[32] Sweden had already banned the use of DDT.

On November 12, 1969, the federal government made a new move. Department of Health, Education, and Welfare (HEW) Secretary Robert Finch announced that, following the recommendation of his Commission on Pesticides and the Human Environment (the Mrak Commission), the government had decided to phase out all "nonessential" uses of DDT over a period of two years.[33] He gave no details.

A week after Finch's announcement, Secretary of Agriculture Hardin issued notices of cancellation of registration for almost all uses of DDT and announced that its use would be almost completely halted by the end of 1971. Simultaneously, the White House announced that the use of other persistent pesticides would be curbed beginning March 1970.[34] These announcements were quite dramatic and made headlines all over the country. It took a little more time for their misleading nature to become apparent.

The USDA had not canceled the registration for use of DDT on cotton, which accounted for two-thirds of its use. Furthermore, the USDA had chosen to "cancel" rather than "suspend" those uses which its order did affect—meaning that manufacturers who appealed the order could continue to sell DDT for the canceled purposes until the appeal was settled. And the appeal process could only have been designed by pesticide industry lawyers;[35] it guaranteed that the appeal could take years to settle. Manufacturers could first ask for the appointment of a special committee of experts to advise Secretary Hardin on whether cancellation was appropriate; if they were dissatisfied with this advice, they could then ask for a public hearing on the matter; and, if they were dissatisfied with the recommendations which came out of the hearing, they could then go to court.[36] Not surprisingly, the manufacturers appealed.

The Environmental Defense Fund Takes the USDA to Court

The EDF and the other conservation groups which had petitioned Hardin to ban DDT were, of course, dissatisfied with the actions which had been taken. At the end of December 1969 they filed suit in the Washington, D.C., U.S. Court of Appeals to have the USDA ordered to *suspend* all the registered uses of DDT.[37] In contrast to cancellation, suspension of registration would have the effect of banning interstate sales of DDT during the appeals process.

Six months later, on May 29, 1970, the Court of Appeals acted on the petition which the environmental groups had filed and ordered Secretary of Agriculture Hardin to suspend the registration of DDT within thirty days or to give reasons for what Chief Judge Bazelon termed "his silent but effective refusal to suspend the registration of DDT." Judge Bazelon dismissed the USDA's cancellation orders as "a few feeble gestures."[38] Hardin responded to Judge Bazelon's order by stating that, in his judgment, DDT did not constitute an imminent hazard to human health, to fish and wildlife, or to the environment, that DDT had essential uses, and that suitable substitutes could not be found for all of these.[39]

The Environmental Protection Agency Takes Over

The USDA was ultimately saved from ever having to cancel or suspend the use of DDT on cotton by President Nixon's creation of the Environmental Protection Agency (EPA). This new agency took over the responsibility for pesticide regulation from the USDA in December 1970.[40]

One of the first orders of business for the head of the new agency, William D. Ruckelshaus, was settling the DDT matter. At the beginning of January 1971, the Court of Appeals—responding again to a petition from the Environmental Defense Fund—ordered the EPA to cancel the registration of *all* products containing DDT and to consider whether the information available to the agency warranted the immediate suspension of registration of these products. The opinion by Chief Judge Bazelon held that cancellation proceedings should be commenced whenever the registration of a pesticide raises "any substantial question of safety," that the secretary of Agriculture had acknowledged that such a question existed in the case of DDT, and that "the statutory scheme contemplates that these questions will be explored in the full light of a public hearing [if requested by a manufacturer], not resolved behind closed doors."[41] A week later the EPA complied with the order to cancel. The manufacturers, of course, immediately appealed the cancellation decision.

Thus, eight years after President Kennedy had ordered the USDA to implement the PSAC recommendation to phase out DDT, the first step had been taken against its major use—on cotton. And the responsibility for this step was taken not by the USDA or its successor agency, the EPA, but by a federal court.

Two months after the cancellation decision, the EPA issued a statement detailing its reasons for not suspending the use of DDT. Perhaps the most substantial reason given was that

precipitous removal of DDT from interstate commerce would force wide-spread resort to highly toxic alternatives in pest control on certain crops. The wide-spread poisonings, both fatal and non-fatal, which may reasonably be projected present an intolerable short-term health hazard.[42]

Although suspension of registration of DDT after eight years would not have been regarded as "precipitous" by many observers, the EPA had raised a legitimate concern. Four fatal poisonings had resulted, for example, when tobacco farmers had switched from the relatively nontoxic DDT to ethyl parathion, a relative of the nerve poisons developed for use in chemical warfare. But in the case of cotton, which accounted by 1970 for 86% of the remaining uses of DDT, the likely substitute, methyl parathion, was already being used mixed with the DDT.[43]

At the end of April 1970, EPA Administrator William D. Ruckelshaus appointed an advisory committee on DDT from a list of nominees provided by the National Academy of Sciences (in accordance with the provisions of the 1947 Federal Insecticide, Fungicide, and Rodenticide Act). Several months later, his DDT advisory committee returned with its report. The committee found that "the evidence to date clearly shows that DDT induces hepatomas and suggests it may be carcinogenic" and that

DDT and its derivatives are serious environmental pollutants and present a substantial threat to the quality of the human environment through widespread damage to some nontarget organisms. There is, therefore, an imminent hazard to human welfare in terms of maintaining healthy desirable flora and fauna in man's environment.[44]

In spite of these findings, the committee did not recommend immediate suspension of the use of DDT, giving as their reason that, even

if one accepts that an eventual health hazard is a possibility, it must be recognized that very little can be done at this time. The world burden of DDT is so high compared to the current annual use in the U.S., that instant as opposed to a rapidly phased cessation of DDT usage would probably make no significant difference in human exposure levels.[45]

The committee's first recommendation was, therefore:

Reduce the use of DDT in the U.S. at the accelerated rate of the past few years with the goal of virtual elimination of any significant additions to the environment.[46]

(The annual domestic use of DDT had declined from a peak of about 70 million pounds in the 1950s and early 1960s to an estimated 12 million pounds in 1970.[47]

Immediately following the release by the EPA of the DDT Advisory Committee's report in September, the Court of Appeals once again ordered the EPA to consider the suspension of all uses of DDT.[48] The cancellation process had already entered its second phase, however: August 17 marked the opening of the hearings which had been requested by some of the manufacturers of DDT, and the EPA decided that it would be "bad policy" to suspend the use of DDT before the hearing process was completed.[49]

It soon became apparent from the conduct of the hearings, however, that both the EPA and the hearing examiner, Civil Service Commission attorney Edmund M. Sweeney, had made up their minds on the issue—and had come to opposite conclusions. The EPA joined with the Environmental Defense Fund to advocate the banning of DDT, while Sweeney's sympathies seemed to be with the pesticide manufacturers, who were joined by the USDA in the defense of DDT. According to *Science* magazine, Sweeney on occasion became quite abusive toward testifying scientists and at one point revealed the extent of his ignorance regarding the proper presentation of scientific evidence by insisting that a scientist answer all technical questions by replying "yes," "no," or "I don't know."[50] Finally, in April 1972, after hearing testimony for seven months, Sweeney announced his conclusion: DDT use did not pose hazards of cancer or birth defects to man and did not "have a deleterious effect on fresh water fish, estuarine organisms, wild birds or other wildlife." He therefore recommended to the EPA that the decision to cancel the registration of DDT be reversed.[51]

Thus, after nearly two years, the cancellation process had neared its end. All that remained was for EPA Administrator Ruckelshaus to make his decision. Speculation abounded regarding the political pressures brought to bear on him. One source inside the EPA's enforcement branch suggested that the fact that most DDT was being used on cotton meant that the decision would be influenced by the Nixon administration's "southern strategy": "This decision is too important to expect the White House to leave it entirely up to the agency."[52] Others, of course, speculated as to what Representative Whitten would do to the EPA budget (over which his subcommittee also had authority) if DDT were canceled. At the same time, the delay in the federal actions on DDT had become rather embarrassing. The states of Washington, Maryland, Wisconsin, and Vermont had joined California, New York, and Michigan in restricting the use of DDT—in most cases by banning its use.[53] And Secretary of the Interior

Walter Hickel had banned the use of DDT on the 500,000 acres of federal land under his control.[54] Internationally, Japan and the Soviet Union[55] had joined Canada, Cyprus, Hungary, Norway, and Sweden in essentially banning the use of DDT.

It came as a welcome surprise to those whom the regulatory history of DDT had taught to be cynical, when finally, on June 14, 1972, Ruckelshaus announced the banning of further DDT use (with some minor exceptions) effective by the end of 1972. The pesticide manufacturers responded by taking the issue to federal court in New Orleans, to which the EDF responded by filing a suit in Washington, D.C., asking that the ban be ruled effective immediately.[56] Both suits were ultimately rejected by the District of Columbia Court of Appeals.[57]

The Significance of the Banning of DDT

The results of the ten-year campaign to ban DDT are open to a range of possible assessments. At one extreme we have the possibility that the campaign against DDT resulted in its being phased out only slightly faster than it would have been as a result of other, natural causes, the most important of these being the widespread development of insect strains resistant to DDT.[58] (The cover of the June 1972 issue of *Environment* magazine showed a large painting of a mosquito with the caption: "As a result of spraying programs, the only thing which will kill this malaria-bearing mosquito is a brick.")

At the other extreme, some consider the banning of DDT as the first step in a worldwide stampede toward the banning of all chemical pesticides. Norman E. Borlaug, winner of the 1970 Nobel Peace Prize for his development of high-yield strains of wheat, expressed this view in November 1971:

DDT is only the first of the dominos. But it is the toughest of all to knock out because of its excellent contribution and safety record. As soon as DDT is successfully banned, there will be a push for the banning of all the chlorinated hydrocarbons, and then, in order, the organic phosphates and carbamate insecticides. Once the task is finished on insecticides they will attack the weed killers and eventually the fungicides.

If the use of pesticides in the U.S.A. were completely banned, crop losses would probably soar to 50 percent and food prices would increase fourfold to fivefold. Who then would provide for the food needs of the low-income groups? Certainly not the privileged environmentalists.[59]

Borlaug testified also to the effectiveness of Rachel Carson's book:

The current vicious, hysterical propaganda campaign against the use of agricultural chemicals, being promoted today by fear-provoking, irresponsible environmentalists, had its genesis in the best-selling, half-science-half-fiction novel *Silent Spring* published in 1962.[60]

And regarding the effectiveness of the environmental groups who had led the campaign against DDT, Borlaug said:

Although the collective membership of these organizations is perhaps less than 150,000, their superb organization and tactics make them an extremely effective force in lobbying for legislation and for brainwashing the public.[61]

Borlaug was obviously upset.

Both these extreme views are probably off the mark. What is clear is that the pesticide manufacturers no longer have unchallenged control over the federal pesticide-regulation machinery. Conversely, the public interest groups have become a force to be reckoned with. By 1971, five years after Yannacone sued the Suffolk County Mosquito Control Commission, the Environmental Defense Fund had grown into a national organization involved in more than a hundred court cases, running the gamut from air pollution to water-resource litigation, with some 32,000 dues-paying supporters and a pool of 700 scientists available as expert witnesses.[62] And the EPA will now respond to reasonable requests from environmental groups without a court order. (Thus, for example, in response to a simple petition from the Environmental Defense Fund, the EPA in March 1971 issued cancellation notices for two more persistent pesticides, aldrin and dieldrin.[63]) At the same time, the record so far hardly supports the view that the environmentalists will soon banish all chemical pesticides. Indeed, some 900 more pesticidal chemicals would have to go through the tortuous cancellation process before Borlaug's nightmare could come true.[64] Even the most hardened environmentalists quail at the thought of such a project.

The story of the struggle over DDT has much to teach those contemplating involvement in efforts to bring about responsible federal policies for technology. Among these is the effectiveness of a well-written book. More than ten years after its publication *Silent Spring* remains a classic influential statement of the case for restraint and care in the use of pesticides—and, by analogy, of technology in general. We have also seen the important options offered to reformers by our federal form of government. Often it is easier to obtain a hearing and mobilize a constituency at a local or state level—on what is really a national issue—than it is to take the issue to the federal government directly. Finally, the Environmental Defense Fund represents an inspirational example of some of the possibilities when scientists and lawyers join forces in the public interest.

NOTES

1. Quoted in James Singer, "DDT Debate Warms Up Again: Should the Government Restrict Its Use?", *National Journal*, November 1, 1969, p. 34.

2. U.S., Executive Office of the President, Office of Science and Technology, *Restoring*

the Quality of the Environment, Report of the President's Science Advisory Committee (Washington, D.C.: Government Printing Office, November 1965).

3. See e.g. Frank Graham, Jr., *Since Silent Spring* (Boston: Houghton-Mifflin Co., 1970), Chapter 7: "Miss Carson's 'Nightmares' Unfold."

4. For a brief summary of the information on the ecological impacts of DDT and other chlorinated hydrocarbons see, National Academy of Sciences, Committee on Oceanography, *Chlorinated Hydrocarbons in the Marine Environment* (Washington, D.C.: National Academy of Sciences, 1971), pp. 8-14. See also U.S., Department of Health, Education, and Welfare, *Report of the Secretary's Commission on Pesticides and Their Relationship to Environmental Health* (Washington, D.C.: Government Printing Office, December 1969), pp. 206-213.

5. Singer, "DDT Debate Warms Up Again: Should the Government Restrict its Use?", p. 34.

6. E. W. Kenworthy, "Full DDT Ban is Refused Pending Review of Safety," *New York Times*, March 19, 1971, p. 1.

7. See e.g. Nick Kotz, "Jamie Whitten: the Permanent Secretary of Agriculture," *Washington Monthly*, October 1969, p. 6.

8. Francis X. Clines, "Suit to Ban DDT in Suffolk County is Dismissed," *New York Times*, December 2, 1967, p. 79.

9. Frank Graham, *Since Silent Spring*, pp. 228-229.

10. Quoted in Jerry M. Flint, "Pesticide Fought in Michigan Suit," *New York Times*, November 12, 1967, p. 79.

11. Quoted in the book on the Wisconsin hearings by Harmon Henkin, Marin Merta, and James Staples, *The Environment, the Establishment, and the Law* (Boston: Houghton Mifflin, 1971), p. vii. See also Luther J. Carter, "DDT: the Critics Attempt to Ban Its Use in Wisconsin," *Science* 163 (1969): 548.

12. Henkin et al., *The Environment, the Establishment, and the Law*, pp. 205-206.

13. See, e.g. Hal Higdon, "Obituary of DDT in Michigan," *New York Times Magazine*, July 6, 1969, p. 36.

14. Ibid., p. 6.

15. Ibid., p. 36.

16. *New York Times*, April 17, 1969, p. 1.

17. *New York Times*, July 17, 1969, p. 50.

18. U.S. Congress, General Accounting Office, Report to Congress on "Need to Improve Regulatory Enforcement Procedures Involving Pesticides" (B-133192, Sept. 10, 1968). Quote is from summary of the report.

19. Quoted in *New York Times*, September 17, 1968, p. 24.

20. U.S. Congress, House, Committee on Government Operations Report, *Deficiencies in the Administration of Federal Insecticide, Fungicide, and Rodenticide Act*, 91st Cong., 1st Session, November 1969, p. 35. See also James Singer, "Recommended DDT Ban May Widen HEW and Interior Regulatory Power," *National Journal*, November 15, 1969, p. 122.

21. U.S. Congress, General Accounting Office, Report to Congress on "Need to Resolve Questions of Safety Involving Certain Registered Uses of Lindane Pesticide Pellets," GAO Report No. B-133192 (Washington, D.C.: GAO, February 20, 1969).

22. *Deficiencies in the Administration of Federal Insecticide, Fungicide, and Rodenticide Act*, p. 35. It is not clear that this number obtained from the USDA is accurate. The Congressional committee notes on p. 36 that "USDA figures showed more products registered over HEW objections in some years than the number of objections actually made."

23. James Turner, *The Chemical Feast, The Ralph Nader Study Group Report on the Food and Drug Administration* (New York: Grossman, 1970), p. 146.

24. *New York Times*, June 17, 1969, p. 42.

25. Ibid., July 23, 1969, p. 47.

26. *Ibid.*, April 18, 1969, p. 86.

27. National Academy of Sciences, *Report of Committee on Persistent Pesticides, Division of Biology and Agriculture to U.S. Department of Agriculture* (Washington, D.C.: NAS, May 1969), p. 29.

28. *New York Times*, August 16, 1969, p. 27.

29. Michael Barone, Grant Ujitusa, and Douglas Matthews, *Almanac of American Politics* (Boston: Gambit, 1972), p. 417.

30. *New York Times*, November 1, 1969, p. 30.

31. *Ibid.*, October 29, 1969, p. 25.

32. *Ibid.*, November 4, 1969, p. 10.

33. *Ibid.*, November 13, 1969, p. 1. We have already discussed the origin of the Mrak Commission in Chapter 6.

34. *Ibid.*, November 21, 1969, p. 1.

35. For a discussion of the influence of pesticide industry lobbyists over the development of pesticide control legislation see *Congressional Quarterly Weekly Report*, October 14, 1972, pp. 2637, 2638.

36. The law which laid out this procedure was the Federal Insecticide, Fungicide, and Rodenticide Act of 1947 (7 U.S.C. 135). On October 21, 1972, this act was amended by the Federal Environmental Pesticide Control Act of 1972 (Public Law 92-516). Among the changes is one which should shorten the cancellation procedure somewhat. Instead of having an advisory committee report and then a public hearing, under the new law, the manufacturers can request a hearing and at any point during the proceeding the hearing officer can request that a committee be set up by the National Academy of Sciences to resolve specific "questions of scientific fact." This committee is given a maximum of 60 days within which to report back.

37. *New York Times*, June 1, 1970, p. 20.

38. Environmental Defense Fund v. Hardin, 428 F. 2d 1093 (C.A. D.C. 1970). See also Jamie Heard, "Chemical Industry, Farmers Fear Pending Pesticide Control Shift," *National Journal*, July 4, 1970, p. 1430.

39. *New York Times*, June 30, 1970, p. 33.

40. *New York Times*, June 6, 1970, p. 21.

41. Environmental Defense Fund v. Ruckelshaus, 439 F.2d 584 (1971) (C.A.D.C. 1971), pp. 593, 594. See also *New York Times*, January 8, 1971, p. 1.

42. U.S., Environmental Protection Agency, *Reasons Underlying the Registration Decisions Concerning Products Containing DDT, 2,4,5-T, Aldrin, and Dieldrin,* (March 18, 1971), p. 16.

43. U.S., Environmental Protection Agency, *Opinion of the Administrator (Consolidated DDT Hearings),* June 2, 1972, pp. 2, 19, 37.

44. U.S., Environmental Protection Agency, *Report of the DDT Advisory Committee,* September 9, 1971, pp. 28, 43.

45. Ibid., p. 28.

46. Ibid., p. 41.

47. For a plot of U.S. DDT production and estimated domestic use through 1969 see Ibid., p. 6. Estimated domestic usage in 1970 and a break-down by major uses are given in *Opinion of the Administrator (Consolidated DDT Hearings)*, pp. 2,3.

48. *New York Times*, September 23, 1971, p. 32.

49. *New York Times*, August 6, 1971, p. 37; October 26, 1971, p. 28.

50. Robert Gillette, "DDT: in Field and Courtroom a Persistent Pesticide Lives On," *Science* 174 (1971): 1108.

51. U.S., Environmental Protection Agency, *Consolidated DDT Hearing: Hearing Examiner's Recommended Findings, Conclusions, and Orders,* April 25, 1972, pp. 93, 94. See also *New York Times*, April 26, 1972, p. 9.

52. Quoted in Gillette, "DDT: in Field and Courtroom a Persistent Pesticide Lives On," p. 1108.

53. *New York Times*, December 30, 1969, p. 14; January 4, 1970, p. 29; February 2, 1970, p. 27.

54. *New York Times*, June 18, 1970, p. 35.

55. *Time*, April 12, 1971, p. 45; *New York Times*, May 14, 1970, p. 6.

56. Robert Gillette, "DDT: Its Days are Numbered, Except Perhaps in Pepper Fields," *Science* 176 (1972): 1313.

57. *EDF et. al* v. *EPA etc.*, D.C. Cir. Op. 72-1548 etc. (6 suits), December 18, 1973.

58. "Decreasing Use of Organochlorines is Result of Insect Resistance, New Chemicals," *Chemical and Engineering News*, August 9, 1971, p. 17.

59. Quoted in the *New York Times*, November 21, 1971, Section IV, p. 13.

60. *Ibid.*

61. *Ibid.*

62. Gillette, "DDT: Its Days Are Numbered," p. 1314.

63. See, for example, Charles F. Wurster, "Aldrin and Dieldrin," *Environment*, October 1971, p. 33.

64. In 1969 there were 900 pesticidal chemicals registered for use in the United States. See, e.g., the *Report of the Secretary's Commission on Pesticides and Their Relationship to Environmental Health*, p. 46.

CHAPTER 11

Matthew Meselson and the United States Policy on Chemical and Biological Warfare

Matthew Meselson is a slight, soft-spoken professor of biochemistry at Harvard who often seems to be occupying the calm at the center of a hurricane of activity. The scene which greeted one of the authors on an afternoon visit to his laboratory during the spring of 1973 was typical: Meselson's graduate students had congregated for wine, cheese, discussion, and laughter in a room next to his office. One door farther down his secretary—long-haired, bearded, and very efficient—was typing. And Meselson himself was working at a table in his office with a student, Robert Baughman, putting the final touches on a paper between telephone interruptions. Meselson apologized sincerely for the fact that he was still finishing up and invited the visitor to look around the office for a few minutes.

The office had the usual academic complement of bookshelves, but their contents were not restricted to books and journals relating to Meselson's professional interests in molecular biology: there were also loose-leaf binders of press clippings, Congressional hearings, reports, and other material on his second great concern of recent years—chemical and biological warfare (CBW). Around the office there was also considerable evidence of Meselson's effort to pull together the final report of the Herbicide Assessment Commission (HAC) sponsored by the 120,000 member American Association for the Advancement of Science. Meselson had led the HAC on a fact-finding trip to South Vietnam in the summer of 1970.

On the easel in the corner stood a topographic map of South Vietnam overlaid with several transparent plastic sheets. Meselson got up to explain that each sheet corresponded to a particular year and that the thin lines on each sheet showed the defoliation and crop-destruction missions flown that year. One's attention was caught by one large mountain valley, perhaps fifteen miles long, covered by many lines. Meselson explained that the valley was customarily blanketed by antipersonnel bombs just before the slow-flying spray planes flew over on a crop-destruction mission. Although U.S. Army officials had originally told the HAC that the valley was unpopulated, Meselson later identified many dwellings on aerial photographs of the area.[1] More recently Meselson had obtained the Army's official figures indicating a civilian population of 17,000 Montagnard tribesmen in the valley.

On several shelves lay stacks of color photographs which Meselson had taken during the HAC visit to South Vietnam. There were pictures of the primitive Montagnard people, many of the women bare-breasted; pictures of a mangrove forest which had been sprayed with herbicides years before—all that was left now was a mass of small barkless tree trunks jutting crookedly out of the bare earth, a grey wasteland; and then there was an aerial photograph of the rich bright green of a living mangrove forest with the dark channels of a river delta winding through it. In December 1970, a few days before the HAC publicly released its preliminary report accompanied by these photographs (but after they had given the White House a preview) the Nixon administration had announced that the herbicide-spraying operations in South Vietnam would be phased out. But by this time almost 10 percent of the area of South Vietnam had been sprayed.

The work which Meselson and his student, Baughman, were now writing up had been stimulated by a problem that had confronted the HAC almost three years before. Meselson and others were concerned about the levels of 2,3,7,8-tetrachlorodibenzo-p-dioxin (abbreviated TCDD or simply called "dioxin") that may have accumulated in the South Vietnamese food supply. Dioxin, which occurs as a contaminant of the herbicide-defoliant 2,4,5-T (discussed in Chapter 6) is extraordinarily poisonous: it is lethal to guinea pigs at doses of 0.6 parts per billion (10^{-9}) of body weight, and it causes birth defects at even smaller concentrations. (The lethal dose for a rat is fifty times higher, that for man is unknown.) What makes dioxin even more dangerous is the fact that it is chemically relatively stable in the environment and that it tends like DDT to accumulate in fatty tissue. As a result, the effects of small doses of dioxin can be cumulative, and it can concentrate in the food chain—and ultimately in man.

In 1970 standard chemical techniques could detect dioxin in food only in concentrations exceeding ten parts per billion—more than ten times the *lethal* concentration for guinea pigs. Since neither the government agencies responsible for regulating pesticides nor the manufacturers of 2,4,5-T seemed particularly interested in improving these techniques, Meselson and Baughman undertook the task. Now, two years later, they had developed a technique which was about

10,000 times more sensitive than the previous methods used (i.e., capable of detecting one part dioxin per trillion by weight). In their first tests on fish samples that the HAC had brought back frozen from South Vietnam, they found dioxin up to concentrations of 0.8 parts per billion. These findings have caused considerable concern and, at the time of this writing, measurements were being rushed on other samples from Vietnam and elsewhere including the United States where 2,4,5-T is used in popular weed and brush killers.[2]

The development of the dioxinmeasurement technique and even the existence of the HAC itself represent only the most recent episodes in Professor Meselson's long involvement with chemical and biological warfare. That involvement began only a few years after Meselson had become a professional scientist, and it has continued for more than a decade.

The Arms Control and Disarmament Agency

Starting in 1957, his first year out of graduate school, Meselson participated in a series of fundamental experiments on the replication of DNA (deoxyribonucleic acid), the molecule which stores and transmits an individual's genetic "code." In 1960 he was appointed associate professor of biology at Harvard, and four years later he was promoted to full professor. Like a number of other promising young scientists, Meselson was introduced to government advising rather early. Several of his older scientific colleagues were already high-level scientific advisors, and in 1963 one of them, Professor Paul Doty of Harvard's chemistry department, then a member of the President's Science Advisory Committee (PSAC), interested Meselson in consulting for the U.S. Arms Control and Disarmament Agency (ACDA). (The ACDA had been established by President Kennedy to prepare for negotiations on the atmospheric nuclear test-ban treaty of 1963.)

Meselson agreed to spend the summer of 1963 at the ACDA and was assigned to study European nuclear defense problems. He soon realized, however, that he could not hope to contribute much of importance on this tangled subject in a summer's time, so he arranged to study chemical and biological warfare (CBW) instead. It was a subject for which his biological background better suited him, and furthermore one which neither the ACDA nor, as it turned out, any other civilian agency had yet subjected to serious review. The State Department, the Defense Department, and the Central Intelligence Agency all offered Meselson excellent cooperation in his summer study, allowing him access to a great deal of secret information. The Army even conducted Meselson and a Harvard colleague, J. D. Watson, on a tour of its chief biological warfare research center, Fort Detrick in Maryland.[3] (Watson, famous as the codiscoverer of the double-helical structure of DNA, was serving at this time on a PSAC panel studying the technical aspects of CBW.)

What Meselson learned profoundly disturbed him. Civilian officials and the top military leadership had repeatedly yielded to constant pressure from the CBW technologists. A series of policy changes, each one relatively minor, had moved America further and further away from its traditional position—which had been unequivocally articulated by President Franklin Roosevelt in 1943, at a time when his generals were considering the use of chemical warfare against the Japanese:

Use of such weapons has been outlawed by the general opinion of civilized mankind. This country has not used them. . . . I state categorically that we shall under no circumstances resort to the use of such weapons unless they are first used by our enemies.[4]

By 1956 a new United States CBW policy had begun to emerge. In that year's edition of the U.S. Army field manual, *The Law of Land Warfare,* the traditional provision that "gas warfare and bacteriological warfare are employed by the United States against enemy personnel only in retaliation for their use by an enemy" [5] was replaced by the following statement:

The United States is not a party to any treaty, now in force, that prohibits or restricts the use in warfare of toxic or nontoxic gases, or smoke or incendiary materials or of bacteriological warfare.[6]

By 1959, the Army CBW establishment had become so bold as to launch a propaganda campaign featuring speeches by Chemical Corps generals (often under the sponsorship of the American Chemical Society) and pro-CBW newspaper and magazine articles, including one in *Harper's* by Brigadier General J. H. Rothschild, commanding general of the (since-reorganized) Chemical Corps Research and Development Command.[7] The purpose of this campaign was twofold: to obtain public and Congressional support for more funding for CBW research and weapons procurement, and to soften public antipathy toward CBW use in combat. It appears to have succeeded at least in the former objective: during the Kennedy administration, spending for CBW increased more than threefold, reaching $300 million per year by 1964. CBW weaponry was now procured on a massive scale and extensively incorporated into Army training.[8]

Even more ominously, in 1961 the Kennedy administration had given the go-ahead to the use of herbicides for defoliating the jungle and destroying crops in "enemy areas" of South Vietnam. Although poisonous gases were not being used, a firebreak had been crossed—the United States was waging chemical warfare. In 1962, during the Cuban missile crisis, an attack on Cuba with an "incapacitating" biological weapon was seriously considered by military officials as part of a U.S. invasion plan. According to Representative Richard D. McCarthy (D.-N.Y.), the plan advanced to the point where Venezuelan equine encephalomyelitis germ warfare agents were placed aboard planes in preparation for use. Although this agent is not officially classified as "lethal," it has been estimated that more than 1 percent of the exposed population would have died as a result of such an attack.[9]

The tremendous American effort to develop such biological ("germ") weapons was particularly disturbing to Meselson. What he had learned about these weapons during his summer at the Arms Control and Disarmament Agency convinced him that they were undesirable on almost every count. At the most fundamental level, he later asked:

What consideration can be given to moral factors in the conduct of war—society's least moral activity? Widespread restraints against certain forms of human combat may be partly based on instinct and accordingly may be wiser than we know. . . . In the course of [the development of increasingly more powerful weapons], governments and people have come to countenance ever increasing levels of destruction in the pursuit of national objectives. At some point this process must be arrested and then reversed if civilization is to overcome the threat to its existence posed by the application of science to warfare. . . . It would be a backwards step to extend the varieties of violence which we now tolerate to include such hitherto reviled means as chemical and biological warfare.[10]

Meselson has also cited many "practical" objections to biological warfare. Thus, although biological weapons might be cheap, might be most suitable for attacking large populations, and might be most effective in a sudden, surprise attack, these are all characteristics that the United States should *not* desire in weaponry. Since the United States already has an enormous arsenal of nuclear weapons, why encourage developments which would make weapons of mass destruction easily available to the smaller nations or to terrorist groups? Moreover, biological weapons would be largely ineffective as battlefield weapons inasmuch as the disease microorganisms require incubation periods in victims of one or more days before taking effect.

Insofar as deterrence is effective, the use of biological warfare by an enemy against United States armed forces should be deterred by the threat of weapons already in existence. Another argument, that the United States has to proceed with the development of CBW weapons in order to be able to develop defenses against them is unconvincing because it would be impossible to prepare, let alone administer, inoculations or other defenses against all the germs which an enemy might employ in warfare. The best general defense against chemical as well as biological attack would be a respiratory face mask, air conditioning, and, in extreme cases, protective suits—devices that would prevent poisons or microbes from coming in contact with their human targets. And the development of such defenses does not require the development of germ weapons themselves.

Meselson wrote a report for the ACDA which was sharply critical of the developing American CBW policy. But the report seems to have been "filed away someplace and probably forgotten," although it may have encouraged the ACDA to undertake the modest series of studies in CBW disarmament which they began in 1964. Meselson thinks that the extensive use of secret information in his report, which he had hoped would give it added authority, may have instead weakened its impact by decreasing its circulation.

Making CBW a Public Issue

Another science advisor might have let it go at that: report submitted, filed, and forgotten. Meselson did not. Since 1963, Meselson estimates, he has spent at least a quarter of his time on anti-CBW activities, increasing to half in the period from 1969 through 1971. At first, he worked mainly as a continuing consultant to the ACDA and also through the international "Pugwash" meetings of scientists interested in disarmament.[11] He was very concerned during this period lest efforts to publicize American CBW activities have the effect of further interesting foreign nations in CBW, which would in turn greatly increase the difficulty of CBW disarmament. But while he was initially worried about publicizing the United States CBW effort in the process of criticizing it, by 1966 Meselson had changed his mind. By this time the U.S. program of forest defoliation, crop destruction, and battlefield use of tear gas in South Vietnam had become truly massive.[12] And as Meselson later explained:

of all the countries in the world, it is the United States which conspicuously pioneers in this area, whose officers and officials consistently have been saying—at lower levels than the President—that these are the weapons of the future. It's the United States which has had conspicuous and major budgetary increases. And it's the United States which has refrained from giving international assurances that it would not be the first to use these weapons.[13]

Meselson therefore decided to join with John Edsall at Harvard in circulating a petition within the scientific community calling for a comprehensive top-level goverment review of the United States' CBW policy. The petition also called for an end to the use of chemical antipersonnel and anticrop weapons in Vietnam and for the reestablishment of the traditional policy forbidding American initiation of the use of CBW.

The job of circulating the petition and collecting signatures was handled primarily by Meselson and a younger biochemist, Milton Leitenberg. They began by sending it to a number of prominent American scientists whose views lay in the center of the political spectrum, reasoning that once the petition had received the endorsement of moderates, more liberal scientists would hasten to add their support. A preliminary petition was released to the press in September 1966 with the signatures of twenty-two leading scientists, including seven Nobel Prize winners.[14] The attendant publicity and the help of the Federation of American Scientists, which sent letters to its entire membership of 2,500, enabled the sponsors to collect the signatures of some 5,000 scientists by the time the petition was presented to the White House on February 14, 1967.[15]

President Johnson's response to the petition is not recorded. He seems to have ignored it. The Pentagon somewhat later began a review of its CBW policies, but that was scuttled.

The petition did contribute to the growth in the public consciousness of CBW as an issue, however. Around 1967 magazine and newspaper articles began appearing which were both well informed and highly critical of current American CBW policy. These were followed by several books. Seymour Hersh's *Chemical and Biological Warfare: America's Hidden Arsenal,* published in spring 1968, was particularly forceful and well documented and succeeded in raising a considerable furor.[16] Meanwhile, in its own inimitable way, the Army committed a massive blunder that focused more attention on the pernicious possibilities of CBW than the anti-CBW scientists could ever have hoped to arouse by themselves.

On March 13, 1968, a cloud of the lethal, highly persistent nerve gas VX from a test spraying accidentally drifted off from the Army's CBW Dugway Proving Ground in western Utah. Within three days, over 6,000 sheep that had been grazing as far away as forty-five miles from the test location were dead. At first the Army refused to admit that they had even been carrying out tests. As the facts became clearer, however, the Army was forced to admit bit by bit, over a period of fourteen months, that its nerve gas had killed the sheep; and it eventually paid damage claims totaling nearly a million dollars. Finally, having been compelled by an aroused Congressional subcommittee "to tell the truth, the whole truth, and nothing but the truth," Army spokesmen reluctantly ended their denials.[17]

The Nixon Administration Review

By February 1969, just after President Nixon had taken office, the United States' CBW program had become so controversial that both the CBS and NBC television networks screened documentary programs on the issue.[18] Neither Nixon nor the Republican party was identified with the CBW expansion which had occurred during the Kennedy-Johnson administration, so the Nixon administration had the opportunity of reexamining the issues on their merits. CBW opponents renewed their efforts to obtain a thorough high-level policy review. Through Presidential assistant Henry Kissinger, who had been his neighbor in Cambridge, Meselson now had a special avenue of access to the President.

At the same time, Congress was beginning to take an interest in CBW. Meselson received an invitation from the Senate Foreign Relations Committee to "educate" it—as the chairman, Senator J. William Fulbright, put it—on the subject. The Committee met for this purpose on April 30, 1969, in executive (i.e., closed) session. A "sanitized" transcript, which became available in June, showed it to have been a remarkably wide-ranging session.[19]

With this indication of increasing Congressional interest, President Nixon, in June 1969, finally ordered the sweeping review of the nation's CBW policy that Meselson and others had long sought.[20] The review was coordinated by Henry Kissinger's office, which analyzed reports prepared by government offices ranging from PSAC to the Defense Department and placed them before the National Security Council and the President for the final policy decisions.

Although Meselson did not participate directly in this review process, he was very active during this period. He prepared and circulated several papers arguing various CBW issues.[21] In addition, Meselson and Doty organized a major American Academy of Arts and Sciences conference during the summer of 1969 in order to "raise the level of discourse" about CBW, as Meselson puts it. A similar purpose was served by a seminar presented before the National Academy of Sciences in October 1969.

Meanwhile, Congress began to respond to the impact of CBW's recent bad publicity. During the same summer, 1969, the Senate Armed Services Committee decided to eliminate all funds in the fiscal 1970 budget for offensive CBW weapons development.[22] The United Nations also got into the picture when one of its study groups, composed of experts from a number of nations, including the United States, issued a detailed factual report on CBW. On the basis of this report, UN Secretary General U Thant called for a halt to the development and stockpiling of chemical and biological weapons and the elimination of these weapons from the arsenals of all nations. Finally, never one to disappoint, the Army continued to make embarrassing CBW blunders: an accident in Okinawa which led to the revelation that the Army had been storing shells and bombs loaded with nerve gas at bases around the world,[23] careless handling of a massive rail shipment of phosgene poison gas across the country,[24] and plans for an even more massive shipment of extremely dangerous nerve gas bombs (discussed in the next section of this chapter). All these developments kept strong pressure on the Nixon administration during its review of America's CBW policies.

On November 25, 1969, President Nixon announced his decision: the United States would renounce first use of lethal and "incapacitating" chemicals and would completely renounce the use of all methods of biological warfare. He also promised to resubmit to the Senate the 1925 Geneva Protocol banning first use of chemical and biological weapons. (Every major nation but the United States and Japan had ratified this treaty by 1931.)

Three months later, Nixon announced that U.S. renunciation of biological weapons would include "toxins"—biologically produced poisons, like the incredibly potent botulism toxin. The National Security Council review of the status of toxins, which had inadvertently been left unclear in President Nixon's previous announcement, had presented the President with three options:

1. Keep toxins.
2. Keep them if they can be produced synthetically.
3. Renounce toxins completely.

In choosing the third option, Mr Nixon went beyond the recommendations of any of his government advisors, including PSAC. He instead followed the advice of CBW critics like Meselson, who argued that national policy should be guided, not by semantic niceties concerning the difference between chemical and biological weapons, but by the desire for eventual worldwide CBW disarmament. Meselson obviously appreciates this decision and others which President Nixon has made on CBW-related issues, for he claims: "I'm a one issue man and CBW is my issue. As far as CBW is concerned the Nixon Administration has been a very good one." Meselson's activities in 1972, however, showed that he was aware of other issues: he worked in the Presidential campaign of Senator McGovern.

The Army's Nerve-Gas Bombs

After seeking scientific advice from highly qualified people, both within and outside the government, we have tentatively concluded that sea burial would offer the least hazard.[25]

—Acting Assistant Secretary of
the Army Charles L. Poor

In April 1969, Representative McCarthy of New York found out quite by chance that the Army was preparing to ship a large quantity of obsolete poison gas across the country for disposal at sea. The poison gas at the Rocky Mountain Arsenal had become a major issue in nearby Denver as a result of reports prepared by the newly formed Colorado Committee for Environmental Information (see Chapter 12), and the Army decided that the easiest way to placate these irate citizens would be to move the gas. They proposed to send it to New Jersey and load it on old Liberty ships, which were then to be towed out to sea and sunk.

McCarthy's interest in CBW dated from the NBC television "First Tuesday" documentary on chemical and biological warfare which he had watched with his wife two months before. As he relates in his book *The Ultimate Folly,* they were shocked by what they saw.[26] When his wife asked him what he knew about CBW, he had to admit his ignorance. The next day he set out to learn more, and he arranged a Pentagon briefing for himself and a number of other Congressmen on March 4, 1969. But the Army did not seem to understand the nature of McCarthy's interest—they used the briefing as an opportunity to campaign for more funds for CBW and refused to answer McCarthy's questions fully. Ironically, McCarthy could have learned much more the same day at MIT, where March 4 had been set aside, as at several other universities, for open discussions of the misuse of science by the government: Meselson spoke there about CBW.[27]

It was inevitable that Meselson and McCarthy would soon get together. The scientist had for some time been talking to Senators and Representatives, their aides, and even some of their larger contributors, trying to arouse some Congressional interest in a curtailment of American CBW activities. Now McCarthy called Meselson for advice about the shipments of poison gas.

Meselson was slow to get excited. When McCarthy first called, Meselson told him that if the shipment only involved relatively nonvolatile susbtances like mustard gas, there should be little danger if reasonable precautions were taken. Both Meselson and McCarthy became greatly concerned, however, as the full dimensions of the Army's plans became apparent: the shipment was to consist of some 800 railroad cars filled with 27,000 *tons* of poison-gas weaponry from Rocky Mountain Arsenal and other munitions depots, including 12,000 tons of lethal GB nerve-gas bombs, 2,600 tons of leaking GB nerve-gas rockets in concrete and steel "coffins," and 5,000 tons of mustard gas.[28] Each railroad car would carry enough poison gas to wipe out several large cities. Representative McCarthy decided to raise a public alarm.

The disclosures resulted in such a general furor that the Army was immobilized. Army spokesmen announced that the shipment would be delayed pending a full investigation by a National Academy of Sciences (NAS) scientific panel. Frederick Seitz, at the time both president of the NAS and chairman of the Defense Department's top science advisory committee, the Defense Science Board, volunteered the services of the NAS for this purpose. To head the special NAS panel, Seitz appointed the famous Harvard chemist and explosives expert George Kistiakowsky. He also tried to appoint the other members of the panel, but Kistiakowsky, who was NAS vice-president and a former science advisor to President Eisenhower, insisted on appointing his own panel. Matthew Meselson was one of Kistiakowsky's appointees.[29]

As a member of the panel, Meselson visited the Rocky Mountain Arsenal and discovered that the technicians there had already accumulated considerable experience dismantling and detoxifying the nerve-gas bombs and were satisfied that they could handle all the 1.6 million "bomblets." Indeed, investigation disclosed that the Army had previously appointed an advisory committee to look into the disposal of nerve gas and that this committee had recommended that the gas be disposed of on site at the Rocky Mountain Arsenal. Despite this advice, the Army brass had quickly agreed to move the nerve gas when it became an issue in the Denver mayoral election.

The NAS panel confirmed that the fears regarding the Army's plans were well founded: they discovered that an average of fifteen derailments per day in the United States had caused, over five years, some fifty evacuations in urban areas. Eight of these incidents had involved trains carrying munitions, and just that spring an ammunition train carrying Vietnam-bound tear gas and explosives had blown up in Nevada. A helicopter inspection by Kistiakowsky of the Army's proposed train route through New Jersey turned up numerous rail crossings without guard-arms. It also became apparent that the Army's proposed emergency medical preparations—a few medics riding on each train, ready to spring out in their rubber suits at a moment's notice to administer atropine to everyone in sight—were ridiculously inadequate in view of the quantity and rapid toxicity of the nerve gas.

Even after the gas reached port and was loaded aboard ships for disposal at sea, the eastern seaboard would not be out of danger. The Army had already dumped a large quantity of munitions, including some less dangerous gas weapons, as part of its "Operation CHASE" (Cut Holes And Sink 'Em"). But these operations were not totally uneventful: one CHASE ship broke loose while being towed to the intended dumping place, and another blew up only five minutes after sinking—apparently as a result of shifting ammunition. The NAS scientists pointed out, in their meeting with the Army officials, the possibility that the excellent acoustic coupling provided by water could cause a massive simultaneous explosion of the nerve-gas bombs when the ships upended as they started to sink. They also pointed out that heavy equipment which was loose aboard the ships could fall onto the bombs and touch off such a chain reaction. When an Army officer denied that the equipment was loose, Kistiakowsky contradicted him with a photograph he had taken only a few days before. If a major explosion of the nerve bombs were actually to occur, the resulting cloud of lethal gas could possibly be carried by the prevailing winds the hundred miles separating the proposed dumping site from New York City. Even slow seepage of the gas would poison a considerable volume of ocean.

The NAS report was released on June 25, 1969. Two days later the Army announced that it had agreed to burn the mustard gas and detoxify and dispose of the nerve gas bombs at Rocky Mountain Arsenal, as the report recommended, rather than shipping them across country.[30] The leaking nerve gas rockets could have been disassembled before they were embedded in concrete, but there now seemed to be no quick and safe method of disposal. They were eventually dumped at sea off the Florida coast. The saga of the Army's surplus poison gas then appeared to be over. But in June 1973 Denver's mayor discovered, in inquiring in Washington why the Army had reneged on its offer to give the city land from the Rocky Mountain Arsenal for a new runway, that disposal of the arsenal's nerve gas had not even begun. Again confronted with outraged citizens, the Army promised to begin destroying the gas in October 1973.[31]

The Herbicide Assessment Commission

We have considered the possibility that the use of herbicides and defoliants might cause short or long term ecological impacts in the areas concerned. ... Qualified scientists, both inside and outside our Government, ... have judged that seriously adverse consequences will not occur.[32]

—John S. Foster, Jr., Director of
Defense Research and Engineering,
September 1967

By 1966 the United States' use of herbicides for defoliation and crop destruction in South Vietnam had reached such a level (about a million acres annually) that many scientists in the United States were moved to protest. In June 1966 E. W. Pfeiffer, Associate Professor of Zoology at the University of Montana, submitted a resolution to the Pacific Division of the American Association for the Advancement of Science (AAAS):

> Whereas units of the U.S. Department of Defense have used . . . [chemical] warfare agents . . . in operations against enemy forces in Vietnam; and
>
> Whereas, the effect of these agents upon biological systems in warfare is not known . . . [and] the scientific community has a responsibility to be fully informed of these agents and their use in warfare because they are a result of scientific research: Therefore be it
>
> *Resolved,* That—
>
> 1. The Pacific division of the AAAS establish a committee of experts in the field of chemical warfare to study the use of CW [chemical warfare] . . . agents in Vietnam with the purpose of determining what agents have been used, the extent of their use, and the effects on all biological systems that might have been affected.
>
> 2. That the above committee make a public report of their findings at the next meeting of the Pacific division of the AAAS.[33]

Pfeiffer's resolution was referred—without recommendation—to the national office of the AAAS.

At its December 1966 meeting, the AAAS Council responded to Pfeiffer's initiative by passing a resolution expressing its concern about the "impact of the uses of biological and chemical agents to modify the environment, whether for peaceful or military purposes," and established a committee "to study such use."[34] Leaning over backward in order to avoid the appearance of entering into the political debate over Vietnam, the AAAS Council broadened Pfeiffer's resolution to the point where the committee which had been created had virtually no instructions at all.

Three months later the committee (to which Pfeiffer had been appointed) came back with the recommendation that the AAAS set up a continuing "Commission on the Consequences of Environmental Alteration" and that various studies be initiated. Vietnam was mentioned as among "areas where massive programs are in progress" and where, the committee suggested, studies of the effects of defoliants might be valuable. But the only suggestion of who might undertake the suggested studies referred to the National Academy of Sciences. Pfeiffer submitted a minority report opposing this suggestion because of the Pentagon's use of the NAS "as a source of advice for biological warfare effort" and also because of NAS's sponsorship of a postdoctoral research fellowship program at Fort Detrick, the Army's main biological warfare research center.[35]

In September 1967 the AAAS sent a letter to Secretary of Defense McNamara suggesting a study of the consequences of the U.S. defoliation program in South Vietnam by either the NAS-NRC, a panel of the President's

Science Advisory Committee, or an independent commission responsible to the Secretary of Defense. The letter was answered by Director of Defense Research and Engineering John S. Foster, Jr., who reassured the AAAS that

qualified scientists, both inside and outside our Government, and in the governments of other nations, have judged that seriously adverse consequences will not occur. Unless we had confidence in these judgements, we would not continue to employ these materials.[36]

But when the president of the AAAS wrote back asking for more information on the technical basis for Foster's "confidence," the Director of Defense Research and Engineering was quite vague, referring only to a "consensus of informed opinion" of fifty to seventy individuals in the absence of "hard data."[37]

Adding to the assurances of his first letter, Foster said that he had commissioned "a leading nonprofit research insititute to thoroughly review and assess all current data in this field" and that he had requested the National Academy of Sciences' National Research Council to set up a panel to "review the results of the study and to make appropriate recommendations concerning it."[38] Four months later, the Midwest Research Institute (MRI), under Department of Defense contract, had reviewed and summarized the literature on the ecological impact of the defoliation program on South Vietnam, and their report had in turn been reviewed by an NAS review panel. The NAS review concluded that the MRI report had adequately surveyed the abundant data on techniques of herbicide use in "vegetation management," adding:

However, the scientific literature provides markedly less factual information on the ecological consequences of herbicide use and particularly of repeated and heavy herbicide applications.[39]

The President of the NAS commented: "Some research in this area is now under way but much more needs to be done."[40] Thus in January 1968, eighteen months after Pfeiffer had asked for a study of the ecological impact of defoliation on South Vietnam because "the effect of these agents upon biological systems in warfare is not known," an NAS panel had reviewed a 369-page summary of 1,500 references and interviews with 147 persons—and had come to essentially the same conclusion.

It seemed to Pfeiffer that it was time for the AAAS to act on his original recommendation. He asked somewhat plaintively:

Are American scientists capable of making an independent study or not? So far the situation has been up in the air. You cannot get the AAAS board of directors to commit themselves to such a study, and I don't think the average AAAS member knows that the study was ever being considered.[41]

Pfeiffer expressed the hope that the AAAS should at least sponsor an extensive symposium on the subject, which "would hopefully stimulate people to go into the field and get data on the effects of herbicides."[42] But, six months later (July 1968), after examining the MRI report, the AAAS Board of Directors again passed the buck by publicly issuing the recommendation that a field study be undertaken under the auspices and direction of the United Nations, with the participation of Vietnamese scientists and scientists from other countries, and with cooperation, support, and protection provided by the contending forces in the area.[43]

This recommendation was sent to the Secretary General of the UN and to the U.S. Secretaries of State and Defense.

The response from the UN was a letter assuring the AAAS that the Secretary General was giving "the matter of chemical and bacteriological weapons . . . his very close attention."[44] The State Department replied that

such studies in combat areas are obviously difficult at present. The United States will be happy to cooperate in responsible long-term investigations of this type as soon as practicable.[45]

And John Foster replied for the Defense Department:

We have continued to gather data and reevaluate all available data and technical judgements. While there are a number of scientific questions left unanswered by available studies, these questions apparently would not be answered by additional, short-term investigations. On balance, we continue to be confident that the controlled use of herbicides will have no long-term ecological impacts inimical to the people and interests of South Vietnam.[46]

Two months later (September 1968), Ellsworth Bunker, U.S. Ambassador to South Vietnam, released the findings of an interagency committee which had reviewed the U.S. defoliation operation. Most of the statements in the report were vague, reflecting a continuing absence of hard data on either the military usefulness or the environmental impact of defoliation. At the end of the report, however, murky and unsubstantiated statements gave way to a very specific conclusion:

Thus, in weighing the overall costs, problems, and unknowns of the herbicide programs against the benefits, the committee concluded that the latter outweigh the former and that the programs should be continued.[47]

The AAAS had thus exhausted the last alternative to taking its own initiative. In December 1968 the AAAS Council finally directed

the AAAS staff to convene, as soon as possible, an ad hoc group involving representation of interested national and international scientific organizations to prepare specific plans for conduct of . . . a field study with the expectation that the AAAS would participate in such a study within the reasonable limits of its resources.[48]

Two-and-one-half years had now passed since Pfeiffer had first submitted his resolution, and over 3 million additional acres of South Vietnam had been sprayed with herbicides. Nothing significant was done during the next year, however, to implement the Council's directive.

Pfeiffer is not one to be stopped easily. Meselson describes him as "a real pioneer type—if he sees a problem, he follows through and explores it wherever it may lead." Pfeiffer decided to undertake an expedition to Vietnam himself. He announced that he and another zoologist, Professor G. H. Orians, would voluntarily conduct a preliminary herbicide assessment expedition to Vietnam under the sponsorship of the small Society for Social Responsibility in Science. Among the objectives of the mission were

to stimulate awareness among scientists of the need for an intensive and long-term study of the effects of military uses of chemical agents in Vietnam [and] to demonstrate the possibility of obtaining meaningful information even with limited funds and personnel. [49]

The expedition was conducted during the second half of March 1969.

In December 1969, the AAAS finally committed itself to action by appropriating $50,000 to fund a Herbicide Assessment Commission which would go to Vietnam to make a pilot study of the environmental and health impact of the defoliation program. Matthew Meselson was invited to organize the study.

Meselson hired Arthur H. Westing, an expert on forest ecology from Windham College in Vermont, as director of the HAC. Both men then surveyed the literature and circulated a proposed list of study topics to over 200 scientists. In June 1970 a five-day working conference at Woods Hole, Massachusetts, attended by twenty-three specialists in such fields as tropical ecology and forestry, helped further to define specific problems for systematic study. Finally, in August and September 1970, Meselson and Westing made a five-week tour of South Vietnam, accompanied by John D. Constable, Professor of Surgery at the Harvard Medical School, and Robert E. Cook, a graduate student in biology at Yale. Constable had already been to South Vietnam representing a Boston-based group called the Physicians for Social Responsibility, which intended to bring severely burned Vietnamese youngsters back to the United States for treatment. This group had received the impression from newspaper reports that many children had suffered burns as a result of U.S. napalm attacks and had survived. But when Constable returned he had to report that he had been able to find very few such victims in the South Vietnamese hospitals that he visited. Meselson was impressed: here was a man who had gone to Vietnam expecting to find something, hadn't found it—and was honest enough to admit as much to the newspapers when he came back. Meselson invited Constable to join the HAC.

Without the cooperation of U.S. and South Vietnamese officials, Meselson and his group could not expect to accomplish much in South Vietnam. Before the HAC left, therefore, the AAAS wrote to Secretary of Defense Laird and to the State Department's Agency for International Development (AID) asking for their cooperation. The response from AID was generous: the group was offered lodgings, food, ground transport, and office facilities while in South Vietnam. But the cooperation sought from the Pentagon was more important—and it was not forthcoming: requests for the locations and dates of herbicide spraying missions were brusquely refused and attempts made in Washington to obtain helicopter transport to sprayed areas were unsuccessful.

The HAC thus arrived in Vietnam armed only with the hospitality of AID and with a letter addressed "To Whom It May Concern" from H. Bentley Glass, Chairman of the Board of the AAAS. It was obvious that the average U.S. official or military officer in South Vietnam was unlikely to be much impressed by such a letter, and it certainly would not get the HAC a helicopter. Meselson therefore began by visiting the U.S. Embassy and the office of the South Vietnamese Prime Minister. When he emerged he had letters of introduction that could be expected to carry some weight.

The HAC's first helicopter ride was obtained by using press cards which had been provided to the group by *Science,* a weekly journal published by the AAAS. But this seemed too much like false pretenses, so they did not use the press cards again. Their next helicopter rides were obtained through the courtesy of the U.S. Embassy—but the Embassy's own access to helicopter transport was so limited that they soon turned elsewhere. When they finally went to the South Vietnamese Army, the letter from the Prime Minister got them complete cooperation: the Vietnamese were willing to order unlimited amounts of helicopter transport for Meselson and his colleagues—from the U.S. Army.

In the meantime Meselson had written to General W. B. Rosson, acting commander of U.S. forces in South Vietnam, renewing his request for information about U.S. herbicide operations in South Vietnam, for helicopter transport, for "logistic and security support to conduct one or two ground inspections," and for statistics recently gathered by the U.S. Army on the incidence of stillbirths and birth defects in South Vietnam.[50] The last item on Meselson's list referred to a study that had been initiated following the release of the Bionetics Research Laboratory study (funded by the U.S. Department of Health, Education, and Welfare) indicating that the herbicide 2,4,5-T is a teratogen.

General Rosson replied that the information on herbicide targets and birth defects Meselson was asking for was classified but that he would be glad to provide helicopter transport. The HAC found this offer virtually unrestricted; they had only to put in a call to get a helicopter whose pilot had orders to "fly as directed" by Meselson, subject only to limitations of safety. The HAC also had access to airplanes belonging to the Vietnam rubber growers' association, whose headquarters in Paris Meselson and Westing had visited on the way to South Vietnam. These airplanes had the advantage that the Vietcong knew them and would not shoot at them; but they were much more difficult than helicopters to take aerial photographs from, so the HAC stuck mainly with the helicopters.

Many of Meselson's flights were with Professor Pham-hoang Ho, a professor of botany who also happened to be South Vietnam's Minister of Education. (Later, after the HAC's report helped bring about the end of the U.S. defoliation program, Professor Ho dedicated his book on the flora of Vietnam to Meselson.) The second in command of the U.S. Chemical Corps in Vietnam also accompanied them. Meselson thought that the Army should be familiar with how the HAC had worked and know the basis for its ultimate conclusions.

On the ground, South Vietnamese professors and students of medicine and zoology helped the HAC collect samples of plants, fish, hair, mother's milk, and so on. The samples were immediately frozen in a 200-pound container of liquid nitrogen. The HAC also recorded interviews with sixty farmers and village officials in or near defoliated areas, including two Montagnard villages.

Although the Pentagon had been uncooperative, the HAC found American military officers in Vietnam generally friendly and open. The HAC did not need very much guidance to find defoliated areas, however. South Vietnam is not a very large country if you have a helicopter, and the defoliated areas were always distinguishable by the dead trees that they contained—the enormous doses of herbicides had not only defoliated but killed millions of trees.

The morning that the Herbicide Assessment Commission left South Vietnam, Meselson had an appointment with General Creighton Abrams, Commander of U.S. forces in Vietnam, who had just returned to duty after undergoing surgery. The interview lasted the entire morning, and Meselson obtained the definite impression that Abrams did not think very much of herbicide use. This impression was confirmed the following December when the *Washington Post* obtained a copy of a cable that General Abrams and Ambassador Bunker had sent jointly to Washington requesting permission to terminate the crop-destruction program.[51] A questionnaire distributed later by the Chief of Army Engineers to officers who had observed the results of defoliation operations in South Vietnam revealed a similar lack of enthusiasm. The responses averaged out to the conclusion that the value of herbicides had been "slight."[52]

When the HAC returned to the United States, the process of analysis and report writing began—and was still going on three years later. Meselson was as creative as usual in obtaining assistance in analyzing the samples he had brought back from Vietnam. For example, since one of the herbicides used for crop destruction, cacodylic acid, is over 50 percent arsenic, it was natural to ask whether it had caused any arsenic poisoning. Meselson got help both from the Boston Metropolitan District Police and from MIT nuclear physicist Lee Grodzins in measuring trace amounts of arsenic in the samples of human hair which the HAC had collected.

The Herbicide Assessment Commission gave a preliminary report on its findings at the annual meeting of the AAAS in Chicago in December 1970. In brief, their findings were as follows:

• • • About half the area of South Vietnam's coastal mangrove forests had been sprayed. U.S. Agriculture Department botanist Dr. Fred S. Tschirley had previously reported that mangroves are killed by herbicide spraying. The pictures that Meselson showed of the lifelessness of these areas years after the spraying gave ample confirmation of this observation. These photographs were widely reproduced in the press and had perhaps the greatest public impact of any item reported by the HAC.

• • • About 20 percent of South Vietnam's relatively mature hardwood forest—which covers almost one-half the area of South Vietnam—had been treated with herbicides, a third of it more than once. Dr. Barry Flamm, chief of the AID Forestry Branch, had previously concluded that a single spraying causes 10 to 20 percent killing of marketable trees, and successive treatments 50 to 100 percent mortality.

• • • A considerable fraction of the crop land in South Vietnam's extensive highlands had been sprayed. These highlands support a population of about a million persons—Montagnard tribesmen—at a subsistence level.

• • • The Commission found some evidence linking the defoliation program with increases in the prevalence of still births in rural Vietnam, but in view of all the war-related disruptions and other factors which might have affected the reported numbers, the evidence did not appear conclusive. The HAC therefore urged further study.[53]

Two weeks before this public presentation, the HAC had given briefings on its findings at both the State Department and the White House (the Defense Department had declined the offer). This was followed, on the opening day of the AAAS meeting, by a surprise announcement from the White House of "an orderly, yet rapid, phaseout of the herbicide operations."[54] We can only speculate on the reasons for this move. But anticipation of the public's revulsion at the vast destruction of Vietnamese forests and food crops must have contributed. At the same time, the request from General Abrams and Ambassador Bunker for an end to the crop-destruction program, along with a general lack of enthusiasm for the defoliation program among Army officers in Vietnam, ought to have made the decision a relatively easy one to make.

The most recent development coming out of the HAC's work—the discovery that dioxin had indeed accumulated in the South Vietnamese food chain—has already been mentioned at the beginning of this chapter. Another development was that Congress ordered in its Military Procurement Authorization Act for the fiscal year 1971 that the Secretary of Defense

undertake to enter into appropriate arrangements with the National Academy of Sciences to conduct a comprehensive study and investigation to determine (A) the ecological and physiological dangers inherent in the use of herbicides, and (B) the ecological and physiological effects of the defoliation program carried out by the Department of Defense in South Vietnam.[55]

Congress asked in the same legislation that the NAS report be submitted by January 31, 1972, but the NAS asked for and received two extensions from the Secretary of Defense and the chairmen of the House and Senate Armed Services Committees.

When the NAS report finally came out in January 1974, it confirmed the seriousness of a number of herbicide effects: reports of illness and death—especially among Montagnard children—following exposure to herbicides; the destroyed mangrove forests would probably take about 100 years to regenerate, they had been invaded by malaria-bearing mosquitos, and the productivity of their offshore fishing grounds had been reduced; defoliation and crop destruction operations had so reduced food supplies in some areas that they had "resulted in the displacement of people from their homes and had contributed to the urbanization of South Vietnam"[56]; and finally the report observed that in South Vietnamese cities herbicides had come to be seen as "an emotionally charged symbol standing for many apprehensions and distresses, especially those for which Americans are blamed."[57] Meselson served on the NAS Report Review Committee panel which reviewed the herbicide report and improved it substantially.[58]

Some Observations

Meselson feels very strongly that the battle against chemical and biological warfare is an all-or-nothing affair. Unless the United States joins with the other nations of the world in ratifying the Geneva Protocol of 1925 which outlaws CBW, he feels that all the successes in the struggle against CBW will soon be forgotten and the whole battle will in a few years have to be fought once again.

Of course, many scientists besides Meselson have played an important role in the opposition to chemical and biological weapons. If we have emphasized Meselson's contributions, we have done so in order to show how effective a single individual can be and how useful it is to be flexible in tactics.

Meselson gained his initial acquaintance with CBW as an "insider," and he has continued to have access to secret data as an advisor to the Arms Control and Disarmament Agency. He has never made public classified information; rather, his clearance enabled him to make sure that his arguments could not be refuted by secret information and established his competence and "credentials" inside the government as well as outside.

Meselson has consistently utilized the advantages of both "insider" and "outsider" positions with remarkable success. Acting in the manner of an insider, he helped the Army make a wise decision on the disposal of its nerve gas, and later he was influential during the Nixon administration's CBW policy review. As an outsider he helped to force first the termination of 2,4,5-T use in Vietnam and later the ending of the entire defoliation and crop-destruction program there. He has also helped to educate Congress and to create and inform the scientific community and popular constituency without whose continuing pressure the "insider" successes would not have been possible. Perhaps most noteworthy of all, in his entire career as an anti-CBW activist Meselson has compromised neither his "future effectiveness" nor his personal scientific integrity.

NOTES

1. This information, and much other material in this chapter, comes from interviews with Meselson. That there were many dwellings in the valley was pointed out by Meselson and John Constable of the HAC in a letter to Ellsworth Bunker, American Ambassador to South Vietnam, November 12, 1970.

2. Robert Baughman and Matthew Meselson, "An Analytical Method for Detecting Dioxin" (paper presented at the Conference on Dibenzodioxins and Dibenzofurans held by the National Institute of Environmental Health Sciences, Research Triangle Park, North Carolina, April 2, 1973). (To be published in *Environmental Health Perspectives.*) See also "Herbicides: AAAS Study Finds Dioxin in Vietnamese Fish," *Science* 180 (1973): 180.

3. David E. Rosenbaum, *New York Times,* November 26, 1969, p. 17.

4. Quoted in Seymour M. Hersh, *Chemical and Biological Warfare: America's Hidden Arsenal* (Garden City, N. Y.: Doubleday Anchor, 1969; originally published in 1968), p. 18. (This reference is referred to below as *CBW.*)

5. Quoted in ibid., p. 19.

6. Ibid.

7. See Jacquard H. Rothschild, "Germs and Gas: The Weapons Nobody Dares Talk About," *Harper's,* June 1959, p. 8.

8. Hersh, *CBW,* p. 28ff.

9. Richard D. McCarthy, *The Ultimate Folly* (New York: Vintage, 1969), pp. 62, 66.

10. Matthew S. Meselson, review of *Tomorrow's Weapons, Chemical and Biological,* by Jacquard H. Rothschild, *Bulletin of the Atomic Scientists,* October 1964, pp. 35-36.

11. Matthew S. Meselson, "A Proposal to Inhibit the Development of Biological Weapons," *Proceedings of the 14th Pugwash Conference on Science and World Affairs* (April 1965), pp. 297-304. For more information on the Pugwash anti-BW efforts see Virginia Brodine, "Detection of Biological Weapons," *Scientist and Citizen,* August-September 1967, p. 168, and Hersh, *CBW,* p. 265. The "Pugwash" conferences began under the sponsorship of industrialist Cyrus Eaton, and were initially held at Eaton's estate in Pugwash, Nova Scotia—hence the name. CBW was first discussed at length in the Pugwash conferences in 1959, and it became a major focus of discussion at 1965 and subsequent meetings. See also J. Rotblat, *Scientists in the Quest for Peace: A History of the Pugwash Conferences* (Cambridge, Mass.: MIT Press, 1972).

12. For a review of U.S. CBW activities in South Vietnam up to 1967, see Hersh, *CBW,* esp. pp. 123-160, and Arthur W. Galston, "Warfare With Herbicides in Vietnam," in *Patient Earth,* ed. John Harte and Robert H. Socolow (New York: Holt, Rinehart, and Winston, 1971), pp. 136-150.

13. *Harvard Alumni Bulletin,* March 11, 1967, pp. 16-19, 30.

14. *New York Times,* September 20, 1966, p. 1.

15. *New York Times,* February 15, 1967, p. 1.

16. Hersh's book (footnote 4) contained a great deal of previously classified information. Hersh had told Meselson that he was studying CBW, but he did not attempt to obtain any information from Meselson and thereby jeopardize Meselson's security clearance.

17. The Dugway incident received considerable attention in the press. The Army's final admission of guilt was reported in the *New York Times,* May 22, 1969, p. 14. A good account is Virginia Brodine, Peter Gaspar, and Albert Pallmann, "The Wind from Dugway," *Environment,* January-February 1969, pp. 2-9, 40-45; reprinted in *Our World in Peril: An Environment Review* (Greenwich, Conn.: Fawcett, 1971), pp. 77-101, Sheldon Novick and Dorothy Cottrell, eds.

18. These television programs may have been inspired by the earlier BBC documentary on CBW: "A Plague on Your Children."

19. U.S. Congress, Senate, Commieee on Foreign Relations, *Chemical and Biological Warfare,* 91st Cong., 1st sess., April 30, 1969 (sanitized and printed June 23, 1969). Excerpts from Meselson's testimony were reprinted in *Bulletin of the Atomic Scientists,* January 1970, pp. 23-24.

20. *New York Times,* June 17, 1969, p. 1.

21. Examples of unpublished papers by Meselson during this period are "The Position of Various Nations During the Interwar Period Regarding the Use in War of Tear Gas Under the Geneva Protocol of 1925" (May 1969), "CS in Vietnam" (July 1969), "The United States and the Geneva Protocol of 1925" (September 1969), and "What Policy for Toxins?" (January 1970).

22. *New York Times,* July 4, 1969, p. 1. The Senate later adopted, by a 91-0 vote, a measure placing numerous restrictions on development, transportation, and storage of CBW munitions (*New York Times,* August 12, 1969, p. 1).

23. *New York Times,* July 19, 1969, p. 1.

24. *New York Times,* August 16, 1969, p. 1; September 9, 1969, p. 33; September 10, 1969, p. 44.

25. Quoted in McCarthy, *The Ultimate Folly,* p. 105, from Mr. Poor's testimony before a subcommittee of the House Foreign Affairs Committee, May 1969.

26. Ibid., pp. viii, 126.

27. Matthew Meselson, "Controlling Chemical and Biological Weapons," in Jonathan Allen, ed. *March 4* (Cambridge, Mass.: MIT Press, 1970), pp. 151-160.

28. Phillip M. Boffey, "CBW: Pressure for Control Builds in Congress, International Groups," *Science* 164 (1969): 1376; "Academy Changes Army Gas Dump Plan," ibid., 165 (1969): 45.

29. The information in this and the following three paragraphs is mainly from interviews with Meselson and Kistiakowsky.

30. Phillip M. Boffey, "Academy Changes Army Gas Dump Plan," *Science* 165 (1969): 45.

31. "Hidden Stores of Poison," *Time,* July 23, 1973, pp. 61-2.

32. Letter from Foster to Don Price, AAAS president, quoted in a statement by the AAAS Board of Directors, "On the Use of Herbicides in Vietnam," *Science* 161 (1968): 253.

33. Quoted in U.S. Congress, House, Committee on Science and Astronautics, *Technical Information for Congress,* Report to the Subcommittee on Science, Research and Development, prepared by the Science Policy Research Division, Congressional Research Service, Library of Congress, April 15, 1971, p. 556. This report contains an eighty-page case study of the involvement of the AAAS in the Vietnam herbicide-use controversy. Our discussion of the background to the AAAS decision to set up its Herbicide Assessment Commission is based on this reference.

34. Quoted in ibid., p. 559.

35. Ibid., p. 560.

36. Quoted in ibid., p. 561. (This is part of the same quote which appears at the beginning of this section.)

37. Quoted in ibid., p. 562.

38. Quoted in ibid., p. 562.

39. Quoted in ibid., p. 567.

40. Quoted in ibid., p. 567.

41. Quoted in ibid., p. 570. (From an interview in *Scientific Research,* January 22, 1968, p. 14).

42. Quoted in ibid., p. 569.

43. Quoted in ibid., pp. 571-572.

44. Quoted in ibid., p. 573.

45. Quoted in ibid., p. 574.

46. Quoted in ibid., p. 574.

47. Quoted in ibid., p. 576.

48. Quoted in ibid., p. 581.

49. Quoted in ibid., p. 585.

50. Letter from Matthew Meselson to General W. B. Rosson, Deputy Commander, U.S. Military Assistance Command, Vietnam, August 12, 1970. (From interview with Meselson.)

51. *New York Times,* December 17, 1970, p. 13.

52. U.S. Army, Office of the Chief of Army Engineers, *Herbicides and Military Operations,* Engineer Strategic Studies Group, February 1972. The first two volumes are unclassified. The third, which discusses primarily possible future wars in which herbicides would be useful, is classified but was reviewed in "Defoliation: Secret Army Study Urges Use in Future Wars," *Science and Government Report,* August 18, 1972. See also Deborah Shapley, "Herbicides: DOD Study of Viet Use Damns With Faint Praise," *Science* 177 (1972): 776.

53. Matthew S. Meselson, Arthur H. Westing, John D. Constable, and James E. Cook, "Preliminary Report of the Herbicide Assessment Commission," presented at the AAAS annual meeting, Chicago, December 30, 1970; reprinted in *Congressional Record* 118 (1972):S3226-33. See also Phillip Boffey, "Herbicides in Vietnam: AAAS Study Runs into a Military Roadblock," *Science* 170 (1970):42; "Herbicides in Vietnam: AAAS Study Finds Widespread Devastation," ibid., 171 (1972):43.

54. Quoted in *New York Times,* December 27, 1970, p. 5.

55. Quoted in *The Effects of Herbicides in South Vietnam: Summary and Conclusions* (Washington: National Academy of Sciences, 1974), p. vii. The best available figures for the total sprayed areas in South Vietnam are given in this report.

56. Ibid., p. S-12.

57. Ibid., p. S-13.

58. Deborah Shapley, "Herbicides: Academy Finds Damage in Vietnam after a Fight of Its Own," *Science* 183 (1974):1177.

CHAPTER 12

Watching the Federal Government in Colorado: The Colorado Committee for Enviromental Information

The history of the Colorado Committee for Environmental Information provides an excellent illustration of the impact that a public interest science group can have at the state level. The committee was most active during the period 1968-1970, when it initiated and informed major debates in Colorado on the hazards connected with three federal programs: (1) the storage of huge quantities of nerve gas at the Rocky Mountain Arsenal near downtown Denver, (2) the continued operation of Dow Chemical's Rocky Flats Plant outside Denver after a disastrous release of intensely radioactive plutonium smoke from the facility had almost occurred and (3) the developmental tests of a method to stimulate the production of natural gas by underground nuclear explosions.

The Rocky Mountain Arsenal

At the Rocky Mountain Arsenal on the outskirts of Denver, the army has manufactured and stored vast amounts of nerve gas and other war gases; in 1968 this stockpile included more than 20,000 nerve-gas cluster bombs containing about 20 gallons of nerve gas apiece.[1] At the height of the cold war, the commander of the arsenal had bragged to a local newspaper reporter that

the gas from a single bomb the size of a quart fruit jar could kill every living thing within a cubic mile, depending on the wind and weather conditions. . . . A

165

tiny drop of the gas in its liquid form on the back of a man's hand will paralyze his nerves instantly and deaden his brain in a few seconds. Death will follow in 30 seconds.[2]

In the wake of the 1968 Dugway incident—in which nerve gas accidentally released during an army test in Utah killed over 6,000 sheep (see Chapter 11)—a more soothing sort of public relations effort seemed to be called for. An article based on an interview with the current arsenal commander appeared in the *Denver Post* beneath a picture showing steel storage tanks of nerve gas neatly stacked like cordwood in an uncovered pile stretching off into the distance. In the story the commander was quoted to the effect that even if "a plane crashed into the drums with sufficient force to release the liquid, it is believed most of it would be absorbed in the ground. A fire would quickly consume the deadly mist."[3]

To a group of scientists in the university town of Boulder, outside Denver, these reassurances smacked of wishful thinking. These scientists were members of an evening discussion group, the "Crossfield Seminar." Led by Dr. Michael McClintock, a physicist at a National Bureau of Standards laboratory in Boulder, they did some simple calculations of what might happen if a fire did not so obligingly "consume" all of the "deadly mist." It seemed quite plausible to them that, in a hypothetical accident like that described above, perhaps 1 percent of the contents of ten ruptured tanks might be blown 150 feet into the air "by the impact of the crash, the accompanying explosion, and convection due to flames."[4] Then, by comparing to the Dugway incident, they found that the resulting "area of lethality" might extend ten miles or so downwind, i.e., possibly into the heart of Denver. The conclusions that McClintock and his collaborators in the discussion group had arrived at were so fearsome that they felt compelled to make their concerns public. After studying the reports on the Dugway incident and the available literature on chemical-warfare agents and weapons, they wrote up a seven-page memorandum on the situation which they released to the press on August 15, 1968.[5]

The memorandum had a substantial impact, receiving both local and national coverage.[6] After a week's silence, the Army let it be known that it had decided to remove the offending nerve gas to a less populated area.[7] Then the public learned, in May 1969, that the Army's plan was to ship the nerve gas bombs by train across the country for eventual dumping into the Atlantic, and there was a new uproar—this time national—with the sequel which we have already described in Chapter 11.

By early 1969, while the nerve-gas controversy was still approaching its climax, the Crossfield Seminar scientists concluded that the nerve gas episode dramatized a more general problem—the public's lack of access to independent technical advice on the environmental and public health implications of governmental programs. To be sure, this was not a new insight. In particular, in the late 1950s there had been massive efforts by scientists to educate the public about the hazards of fallout from nuclear testing. (These efforts paved the way for negotiation of the Partial Nuclear Test-Ban Treaty of 1963.) Certain organizations which were formed in that struggle became permanent and have continued the effort of public education on issues relating to the impact and control of technology. Among these are the St. Louis-based Committee for Environmental Information, which founded the magazine *Environment,* and the New York-based Scientists' Institute for Public Information (SIPI), which acts as a national umbrella organization for the St. Louis committee and about twenty other science information committees in other parts of the country.

In March 1969 the Colorado group decided to organize itself as the Colorado Committee for Environmental Information (CCEI), a nonprofit corporation affiliated with SIPI.[8]

The CCEI almost immediately found itself embroiled in two new issues: the danger of plutonium contamination of the Denver area resulting from activities at a nearby Atomic Energy Commission (AEC) nuclear weapons fabrication plant, and the danger of radioactive pollution from an AEC-promoted program to increase the production of natural gas from certain Colorado rock formations by fracturing them with underground nuclear explosions.

Plutonium Pollution

On May 11, 1969, a fire in the Dow Chemical Company's Rocky Flats plant, sixteen miles from downtown Denver, caused about $100 million worth of damage.[9] This was not an ordinary factory nor an ordinary fire: the plant, run by Dow for the AEC, makes plutonium nuclear triggers for thermonuclear weapons, and the fire, the largest industrial accident in history, involved about 1,000 pounds of plutonium.[10]

The artificial element plutonium is terribly dangerous in the form of smoke or dust. Less than a millionth of a gram of tiny particles of plutonium oxide lodged in a human lung will intensely irradiate the neighboring tissues with short-ranged alpha particles over a period of years, with lung cancer a likely result. If a significant fraction of the plutonium involved in the Rocky Flats fire had escaped to the outside air, the result might well have been, as the AEC later acknowledged, a public health catastrophe for the entire Denver area.[11] The public was immediately reassured by spokesmen for the AEC, Dow, and the Colorado Department of Health, however, that the plant's air-filtration system had worked effectively during the fire and that there had been no release of plutonium into the atmosphere.

The CCEI group first learned about the fire from a newspaper which one of the scientists brought to their regular meeting the next day. The discussion which ensued quickly focused on two questions: (1) Was it possible that the smoke from such a major fire could really have been contained so effectively? (2) Would it not be tempting fate to continue the Rocky Flats plant in operation so near to a major population center after this near-disaster? A subcommittee was set up to look into these questions under the chairmanship of Dr. E. A. Martell, a nuclear chemist at the National Center for Atmospheric Research in Boulder and a world-recognized expert in the methods of detecting trace amounts of radioactive isotopes in the environment. Two weeks later the CCEI made its concerns public—this time in the form of an open letter to Colorado's Governor John Love:

Since published reports contained no information indicating that an adequate survey has been made of the large areas outside of [the Rocky Flats plant], it is possible that large amounts of toxic plutonium oxide could have been deposited as fallout from the smoke plume miles downwind from the plant.

The wisdom of the AEC in keeping such a facility in the center of the largest metropolitan area between the Missouri River and the West Coast must be seriously questioned.[12]

The letter then went on to list a number of detailed questions concerning the technical basis for the claim that no plutonium had escaped from the plant. The scientists questioned whether either Dow or the Colorado Department of Health had used the specialized equipment necessary to detect plutonium contamination. Copies of the letter were hand-delivered to the media by Peter Metzger, president of the CCEI.

Metzger's dealings with the media deserve a discussion in their own right. A tall, balding, playfully contentious biochemist who at the time was 38 and employed by Ball Brothers, a research laboratory in Boulder, Metzger recalls that when he first began delivering CCEI releases to local newsrooms he was generally regarded with profound suspicion. The tidings he bore were so disturbing that some of his contacts accused him of being a "Communist." It was only when Metzger interested outside newspapers—notably the *New York Times* and the *Los Angeles Times*—in covering CCEI stories that the local media people started to listen too when he came around. Metzger's rounds with each CCEI press release eventually expanded to twenty-three stops, including every newspaper and every television and radio station in Denver. He soon learned that newsmen rapidly lose interest in a story if they feel that they have been or will be scooped. He therefore adjusted the timing on the releases so that the news would come out at about the same time from as many sources as possible. (Ultimately, Metzger enjoyed his dealings with the press so much that he began writing articles on the controversies for the *New York Times Magazine.* He then dropped his career in biochemistry altogether to write a book, *The Atomic Establishment,* and do a weekly column of "science and technology muckraking" for the *New York Times* syndicate.) Before long Colorado newsmen began coming to CCEI for information. The scientists then learned, after one or two bad experiences, that it was important to have a well-informed contact man on each issue. The problem was that the newsmen would tend to go to the CCEI signatory whose name they knew best, but that scientist might not be the best informed on that particular issue. To avoid this, Metzger and Dr. Robert Williams, an energetic and articulate young physicist at the Environmental

Sciences Services Administration Research Laboratories in Boulder, were usually indicated on the CCEI releases as press contacts.

Metzger did his work well: the letter from the CCEI to Governor Love about the Rocky Flats plutonium fire was widely reported in the Colorado press. As might be expected, AEC and Dow spokesmen reassured the public about the extensive observations on which the claims of no plutonium escape were based.[13] But Governor Love called up General Edward B. Giller, director of the AEC's Military Applications Division, to ask him for a briefing on the matter. General Giller in turn called Dr. Martell, whom he knew from an earlier period when they had both been involved in the nuclear weapons testing program in the Pacific. (Martell, a retired Air Force colonel, had been program director of the Armed Forces Special Weapons Project.) Two meetings were arranged for Giller and other AEC and Dow officials—one with the governor and one with the CCEI scientists.

After his briefing Governor Love emerged to report that General Giller had assured him that there was no danger to the public as a result of the Rocky Flats fire. This announcement effectively undercut the CCEI position that the public health should be safeguarded by more than the assurances of the agency whose operation was being questioned. Giller's visit did have some compensations for the CCEI scientists, however: in their meeting with him they were able to exact his commitment to have Dow answer a list of specific technical questions concerning its measurements of plutonium losses from the plant and the extent of contamination of the area surrounding the plant.[14]

The answers to the CCEI questions came back with a key omission: the AEC, Dow, and the Colorado Department of Health had all refused to check soil samples in the area around the Rocky Flats plant for plutonium contamination. They argued that the significance of such samples would be difficult to evaluate and that, anyway, the level of airborne radioactivity was a much more direct measure of the public health hazard.[15]

Fortunately, however, the CCEI had the means for breaking this impasse: Dr. Martell was a master of the delicate techniques required to detect traces of plutonium. Martell therefore undertook an extended program of measurements in his laboratory on more than 100 soil samples taken at various locations from two to ten miles from the Rocky Flats plant. In February 1970, after months of work, he made his results public: at least 1,000 times as much plutonium had escaped from the Rocky Flats plant as could be accounted for by Dow figures for the previous year, including those for the May fire.[16] (Martell's subsequent measurements revealed that most of the excess plutonium in the Rocky Flats area was not due to the May 1969 fire but rather had been released in a series of accidents over a period of years prior to that date.[17]) Meanwhile Giller, having learned of Martell's study, had commissioned a similar soil-sampling program himself; and the results of this study essentially corroborated Martell's findings. (It is amusing to note that Rocky Flats personnel contacted Martell for technical advice on how to do the study.) But the AEC nevertheless insisted that the level of plutonium contamination involved still constituted an insignificant health hazard, while Dow Company spokesmen pointed to upgraded safety features being incorporated into the plant as it was being rebuilt.

The CCEI scientists took advantage of the new burst of public attention resulting from Martell's findings to attempt to communicate once more what they felt were the major issues which should be confronted by the state and federal government. First, they pointed out that there was disagreement within the scientific community about the danger associated with what the AEC considered a "permissible lung burden" of plutonium. Some scientists were arguing that the AEC's level had been set too high by a factor of 100. Second, they raised once again the question of whether the Rocky Flats plant constituted such a public health hazard that it should be relocated away from the Denver area. Martell commented: "We can't afford to wait until we are in trouble, because then Denver will have to move instead of Rocky Flats."[18]

In fact, after Martell's findings were made public there came some very disturbing revelations concerning plutonium-handling practices at Rocky Flats. For example, it seemed that some of the plutonium contamination detected by Martell was due to leakage of contaminated oil onto the ground in a storage area: some of the oil-soaked dirt had dried and blown away.[19] Another revelation following the May 1969 fire was that the Rocky Flats plant had been suffering an average of more than one plutonium fire per month.[20] A CCEI press release commented that while "it is not possible to make realistic predictions about the number and magnitude of plutonium releases in the future, . . . it can only be stated that the record up to now is not very reassuring."[21]

Despite the tumult following the publication of Martell's findings, the issues which the CCEI had raised soon began to fade again unresolved. Governor Love easily beat back the political challenge of Lieutenant Governor Hogan who had tried to make the governor's passive attitude toward the AEC into an election issue; and the state legislature, following the governor's wishes, refused to assert Colorado's right to set safety standards higher than those of the AEC.[22] The public appeared generally willing to accept Dow's assurances that safety-motivated design changes which were being incorporated into the Rocky Flats plant would prevent another major fire. It thus appears that the main effect of the controversy was to make both the AEC and Dow management much more concerned about fire prevention and plutonium-handling practices at Rocky Flats.[23] They were also put on notice that their public relations statements were subject to check by independent scientists.

Nuclear Stimulation of Natural Gas Production

"Plowshare" is the AEC's name for its program for developing peaceful applications of nuclear explosives. One proposal is to liberate natural gas trapped in relatively impervious rock formations by fracturing the rock with such explosives. A test of this method, Project Rulison, was scheduled to take place in Colorado's Rulison natural-gas field in the fall of 1969.

Underground nuclear explosions are no novelty. Since the United States signed the Partial Nuclear Test-Ban Treaty in 1963, the AEC has announced an average of about thirty underground nuclear weapons tests in Nevada each year.[24] However, use of the nuclear gas-stimulation technique in the production of a significant proportion of U.S. natural gas would require many thousands of nuclear explosions.[25] To the CCEI scientists, the environmental impact of such an unprecedented program seemed well worth studying. A subcommittee made up of Metzger, Martell, and Williams was set up to look into the matter.

The CCEI scientists were mainly concerned about the fate of the large amount of radioactivity released in each nuclear explosion. Other potential hazards—landslides, mine cave-ins, bursting dams, falling chimneys, and cracking plaster—would be all too evident to those who lived and worked in an area where nuclear gas stimulation was in progress. But radioactivity is invisible; its health effects, such as cancer and gene damage, are delayed for decades or generations; it might take many decades before the radioactive poisons left underground by the explosions were leached out by water and brought to the surface to contaminate man's food and water. Independent scientists were needed who could evaluate and explain these hazards to the public.

On July 28, 1969, the "Rulison Subcommittee" of the CCEI issued a press release raising "serious questions concerning the potential hazards connected with Project Rulison."[26] They emphasized the magnified hazards which would be associated with the adoption of the nuclear gas-stimulation technique on a large scale. Thus:

If the entire Rulison field is developed by this technique, it will mean that rock beneath 60,000 acres in our state will have been fractured to facilitate the flow of natural gas and that enormous (i.e., megacurie) quantities of strontium-90 and cesium-137 will have been distributed underground. . . . If it were discovered some years later that . . . underground water contamination was occurring, it would be too late to do anything about it.[27]

In response to the CCEI press release, the AEC rushed in once again to reassure Governor Love and the Colorado public. Representatives of the private companies collaborating in the project, the AEC, the U.S. Public Health Service, the Bureau of Mines, the U.S. Geological Survey, the AEC's Los Alamos National Laboratory, and the Colorado Public Health Service all met with Governor Love to impress on him the absence of hazard from the Rulison test. They followed this meeting with a news conference in which the same reassurances were offered the public. Governor Love lent his authority to their message the next day by announcing that he was "certainly . . . impressed by the safety precautions. . . . It's my opinion they have built in a safety factor that is, in all likelihood, greater than will be required. . . . I can find no reason to object on the grounds of safety."[28]

It was now less than a month before the scheduled Rulison blast, and

Governor Love's statement seemed to confirm the impression that the state government was not willing even to explore the possibility of opposing the AEC. The only recourse for opponents of the test, then, appeared to lie in the courts. Metzger had already stirred the interest of the Colorado branch of the American Civil Liberties Union (ACLU) by inviting its representatives to discussions of the matter with CCEI scientists.[29] On August 22, 1969, ACLU lawyers filed a complaint in the Denver U.S. District Court asking for an injunction to stop the test. An environmental group, the Colorado Open Space Coordinating Council, quickly joined in the suit.[30] After hearing the case, in which Metzger and Martell appeared as witnesses along with many AEC experts, Judge Alfred A. Arraj refused to issue the requested injunction against the blast itself on the grounds that the radioactivity resulting from the blast would remain isolated underground until flaring of the released gas began. He left the way open, however, for the plaintiffs to seek another injunction later against the flaring of the gas. The decision was upheld on appeal.[31]

In the meantime the CCEI had partly succeeded in getting the AEC to make public the technical basis for its assertions that the Rulison test and later commercial application of the nuclear gas stimulation method would not result in excessive public health hazards. On August 6 the CCEI scientists had submitted to the AEC a list of detailed questions concerning the types and amounts of the radioactivity which would be created by the blast: How much radioactivity would end up in the gas, in the water, or be trapped in the glasslike rock created by the heat of the explosion? What would be the AEC's criteria for allowable radioactivity in the flared gas and later for gas which would be distributed commercially? What was the distribution of underground faults in the area of the Rulison blast? And what financial liability would the participating corporations and government agencies assume if commercial use of the nuclear gas-stimulation technique resulted in serious damage to or radioactive contamination of the local environment?[32]

No answers had been received to these questions eight days before the scheduled date of the blast, September 4, 1969, when CCEI representatives visited Governor Love, after which Love publicly expressed his interest in hearing the AEC's answers to three specific questions which the CCEI scientists had raised.[33] Two days later the AEC submitted answers to the governor's questions—as well as to many other questions which had been raised by the CCEI.[34] Governor Love seems to have been satisfied by the AEC's answers—but the CCEI was not. As Metzger explained in a letter to Love:

The serious questions raised concerning long-range public health and safety problems have been either ignored or answered unresponsively. . . . There can be no justification for the Rulison shot if the full-scale application of nuclear gas stimulation technology involves unacceptable risks to the public and both serious damage and persistent contamination of the local environment.[35]

On September 10, 1969, after several days' delay because of adverse weather conditions and with helicopters sweeping the area in an attempt to keep protesters away from the site, the Rulison nuclear device was detonated with the force of 40,000 tons of TNT (two Hiroshima-sized bombs) more than a mile-and-a-half beneath the earth's surface.[36] Reporter Cal Queal of the *Denver Post* later collected the following reactions of local residents to the effects on the land above:

> Lannie Dix told what it was really like as he stood on a bluff at Rifle [twelve miles away], looking west at 3 p.m., September 10.
> "You could see the ground swell, just like waves on the sea," he says. "There were three waves—up, then down—and the ground rolled under your feet each time."
> He paused and shook his head. "There's nothing under the ground that's worth that."
> In Grand Valley, 6½ miles from the bomb, Otto Letson sat in his automobile when the shock came.
> "It felt like someone picked up the car about eight inches, shook it, and then set it back down," he said. "Dust came off all those hills and rocks were rolling down everywhere."[37]

The legal battle was immediately renewed as ACLU lawyers, lawyers for the Colorado Open Space Coordinating Council, and the district attorney for Colorado's 9th Judicial District joined in an attempt to obtain an injunction from Judge Arraj barring the AEC and its industrial partners from drilling back to tap the gases which had been freed and made weakly radioactive by the explosion.[38] Although the judge again ruled in favor of the AEC, the concerns expressed by the CCEI about the public health hazards which might result from a massive use of the nuclear gas-stimulation method apparently had had some impact on him. In his opinion, Judge Arraj cautioned:

> Lest our ruling today be misunderstood, some additional words are required. ... We are not here and now approving continued detonations and flaring operations in the Rulison field. Such determination must be made in the context of a specific factual situation, in light of contemporary knowledge of science and medicine of the dangers of radioactivity, at the time such projects are conceived and executed.[39]

Judge Arraj also made legally binding the AEC's previous commitment promptly to make public the data obtained from a rather elaborate system set up to monitor the amount of radioactivity released with the gas from the Rulison field and the extent of accumulation of this radioactivity in the water, vegetation, and milk in the surrounding area.

Thus, while the challengers had not stopped the Rulison test, their efforts had not been without effect. The AEC was put on notice for the first time that the public health hazards of its activities were subject to court review. The public had been alerted to the possible hazards of the nuclear gas-stimulation technique—Colorado editors voted the debate over Project Rulison the state's number-one news issue of the year.[40] And the local press had shown itself to be no longer willing to accept reassuring press releases from the AEC without independent review of the technical facts. It is not clear how seriously the AEC took the opposition to its Rulison test, but in other parts of the government it was taken very seriously. Following the episode, a staff report of the Federal Power Commission's Bureau of Natural Gas, after expressing doubts about the economics of the nuclear gas-stimulation method, made the following comment:

> There are political and long range environmental consequences to be considered. In order to substantially increase natural gas availability, . . . thousands of nuclear devices will have to be detonated. In view of the increasingly forceful and articulate expressions of concern being voiced for the integrity of the natural environment, such large-scale applications might not gain public acceptance.[41]

Conclusion

We have seen how the Colorado Committee for Environmental Information raised questions about the public health hazards of three federal activities in Colorado and thereby triggered intense public controversies. In each case, after the controversy had died down, the situation was substantially changed: the Army had committed itself publicly to the destruction of its nerve-gas stockpiles at the Rocky Mountain Arsenal; plutonium-handling practices at Dow's Rocky Flats plant were much upgraded; and the public acceptability of a large nuclear gas-stimulation program was thrown into considerable doubt. On the other hand: In 1973 the nerve gas was still stored next to Denver's airport, essentially as it was in 1968 when McClintok and his group first raised the issue; Dow's Rocky Flats plant was still there, on the outskirts of Denver, handling huge quantities of extremely dangerous plutonium; and the AEC carried through with the Rulison test, and in May 1973 it conducted another nuclear gas-stimulation experiment ("Rio Blanco") in Colorado.

The history of the CCEI is inspirational in that it demonstrates how a small group of scientists can make accessible to the public—at the state level, at least—technical issues which have serious implications for the public health and welfare but which would otherwise be dealt with behind closed doors—or perhaps even not be dealt with at all. Although the most active members of the CCEI are now dispersed, the committee has left as a legacy in Colorado a much more alert and resourceful news community (enriched to no small extent by the fact that in 1974 Peter Metzger became a full-time newsman for the *Rocky Mountain News*).

One of the more interesting outcomes of the CCEI's activities was its impact on the careers of its leadership. Metzger, McClintok, and Williams have all shifted their careers in the direction of public interest science.

Peter Metzger, as we have mentioned, traded in his career as a research biochemist at an industrial "think tank" for one as a "science and technology muckraking" newsman.

Michael McClintok moved to the University of Wisconsin, where he again became embroiled in a public controversy with the military—as a technical critic of the Navy's Project Sanguine.[42] In 1973 McClintok joined the Program on Technology and Man at Clark University in Worcester, Massachusetts.

Finally, Robert Williams moved to the Department of Physics at the University of Michigan, where his interests took him into energy studies. By 1972 he held a responsible position at the Washington-based Energy Policy Project, funded by the Ford Foundation.

The effects of their participation in the CCEI on these scientists' careers testifies to the excitement such an involvement generates, as well as to the almost irreversible nature of the commitment one makes when he becomes seriously involved in public interest science.

NOTES

1. As of July 1973, the inventory still included 21,115 cluster bombs containing a total of 463,000 gallons of GB nerve gas, 5.5 million pounds of mustard gas (about an equal amount had been destroyed in the previous year), 2.6 million pounds of phosgene gas, and an undisclosed amount of GB nerve gas stored in bulk tanks and unfused bombs. See James P. Sterba, "Nerve Gas to Stay in Denver Area", *New York Times,* July 5, 1973, p. 20.

2. Quoted in Seymour M. Hersh, *Chemical and Biological Warfare—America's Hidden Arsenal* (Garden City, N.Y.: Doubleday Anchor, 1969), p. 90.

3. Dan Partner, "Arsenal Nerve Gas Poses No Danger, Official Says", *Denver Post,* March 28, 1969, p. 40.

4. "Memorandum on the Possible Hazard to Denver of the Rocky Mountain Arsenal's Storage of Nerve Gas," August 15, 1968 (signed by Michael McClintock, Frank Oppenheimer, Jonathan B. Chase, Lester Goldstein, David R. Crosley, George Wm. Curtis, and Lew Trenner). This memorandum, along with other CCEI documents and clippings, was provided to the authors by Robert H. Williams.

5. *Ibid.*

6. See e.g. the *New York Times,* August 18, 1968, p. 27.

7. *New York Times,* September 8, 1968, p. 36.

8. *Boulder Daily Camera,* March 23, 1969, p. 19. A description of the activities of the St. Louis group (then called the St. Louis Committee for Nuclear Information) during the fallout controversy may be found in Barry Commoner, *Science and Survival* (New York: Viking Compass, 1966), pp. 110-120.

9. *Denver Post,* May 12, 1969, p. 3.

10. Our estimate is based on the report by an AEC spokesman (*Denver Post,* June 4, 1969, p. 19) that $20 million worth of plutonium would have to be recovered from the premises and a price of $40 per gram ($1,160 per ounce) for "weapons grade" plutonium.

11. In October 1970 General Edward B. Giller, AEC Assistant General Manager for Military Applications, justified a request for emergency appropriations to upgrade the safety features of the Rocky Flats Plant as follows: "If a major fire were to break out and break through the building, that is breach the roof, then hundreds of square miles could be involved in radiation exposure and involve cleanup at an astronomical cost as well as creating a very intense reaction by the general public exposed to this . . . In the fire we had

last year we kept it in the building. If the fire had been a little bigger it is questionable whether it could have been contained." (U.S., Congress, House Committee on Appropriations Hearings, *Supplemental Appropriation Bill, 1971,* 91st Congress, 2nd Session, October 1, 1970, p. 295). The hazardous nature of plutonium is discussed by Donald Geesaman, "Plutonium and Public Health," Lawrence Livermore Laboratory, Livermore, Calif. Report No. GT-121-70, April 19, 1970. Reprinted in U.S., Congress, Senate, Committee on Public Works Hearings, *Underground Use of Nuclear Energy, Part 2,* 91st Cong., 2nd sess., August 5, 1970, pp. 1524-1537.

12. Excerpts from the letter, dated June 4, 1969, were reprinted in a number of Colorado papers. The full text was reprinted in the June 4, 1969 edition of the Boulder newspaper, *Town and Country.* The letter was signed by E. U. Condon, Robert H. Williams, Michael McClintock, Dion W. J. Shea, Edward A. Martell, George William Curtis, and H. Peter Metzger.

13. Bob Huber, "No Contamination Reported in Rocky Flats Fire," *Denver Post,* June 8, 1969, p. 1.

14. *Denver Post,* June 18, 1969, p. 1.

15. Ken Pearce, "CCEI Pressed Dow to Find Radioactivity", *Denver Post,* February 13, 1970.

16. *New York Times,* February 11, 1970, p. 1.

17. Robert Williams, private communication.

18. Robert Threlkeld, "CCEI Attacks Official Stand on Contamination," *Rocky Mountain News,* February 25, 1970.

19. *Dow Corral,* March 24, 1970. (The *Corral* is the house organ of the Rocky Flats plant.) The purpose of this edition was to "compare charges, suppositions, and conclusions made in a recent CCEI draft report with the facts." (p. 1) On page 6 the *Corral* discusses possible sources of the contamination including the leaking oil drums.

20. Roger Rapoport, "Secrecy and Safety at Rocky Flats," *Los Angeles Times,* September 7, 1969, "West" Section.

21. Robert Threlkeld, "CCEI Attacks Official Stand on Contamination."

22. See e.g. Fred Brown, "Bills to be Offered on Pollution Control," *Denver Post,* February 14, 1970, p. 24.

23. The reasons for the May 1969 fire and the measures taken to prevent its recurrence are presented in AEC reports reprinted as appendices to: U.S. Congress, Joint Committee on Atomic Energy Hearings, *AEC Authorizing Legislation, Fiscal Year 1971,* 91st Congress, 2nd Session, March 19, 1970, Part 4, pp. 1946-1997. For a criticism of the management of Rocky Flats two years after the fire and for additional valuable material on the plutonium contamination controversy see Deborah Shapley, "Rocky Flats: Credibility Gap Widens on Plutonium Safety," *Science* 174 (1971): 569-572 and a letter criticizing the article by Donald E. Michels, *Science* 177 (1972): 208.

24. The total number of U.S. underground nuclear tests announced during the period 1963-1971 is 229. See e.g. Robert Neild and J. P. Ruina, "A Comprehensive Ban on Nuclear Testing", *Science* 175 (1972): 140.

25. The number of blasts to develop just one 93,000 acre gas field in Colorado has been estimated at 1,000: Luther J. Carter, "Rio Blanco: Stimulating Gas and Conflict in Colorado," *Science* 25, 1973: 847.

26. The press release was signed by H. Peter Metzger, Robert H. Williams, and Edward A. Martell. See Bob Huber, "Closer Look at Nuclear Blast Urged", *Denver Post,* July 28, 1969, p. 40.

27. CCEI Press Release, July 28, 1969, p. 3.

28. *Denver Post,* August 13, 1969.

29. *Boulder Daily Camera,* May 15, 1969.

30. *Denver Post,* August 23, 1969, p. 28; see also *New York Times,* August 26, 1969, p. 8.

31. *Rocky Mountain News,* August 28, 1969; *Denver Post,* September 3, 1969, p. 2.

32. CCEI, "Questions Relating to the Forthcoming Rulison Underground Nuclear Explosion in Western Colorado," August 6, 1969. Signed by H. Peter Metzger, Edward A. Martell, Robert H. Williams, and A. Skumanich.

33. *Denver Post,* August 27, 1969.

34. Dick Prouty, *Denver Post,* August 31, 1969.

35. Letter to Governor Love from H. Peter Metzger dated September 5, 1969. See also CCEI press release dated August 31, 1969, quoted in many news stories including *Rocky Mountain News,* September 1, 1969, p. 67.

36. *Rocky Mountain News,* September 11, 1969, p. 1.

37. *Denver Post,* April 26, 1970.

38. A summary of the issues and testimony in this suit may be found in Judge Arraj's opinion, which is reprinted in U.S. Congress, Joint Committee on Atomic Energy Hearings, *AEC Authorizing Legislation, Fiscal Year 1971,* March 3 and 5, 1970, Part 2, pp. 1106-1130.

39. Quoted in *Ibid.,* p. 1128.

40. *Denver Post,* December 26, 1969.

41. U.S. Federal Power Commission, "Staff Report on Natural Gas Supply and Demand", September 1969, quoted in Peter Metzger, "Project Gasbuggy and Catch-85," *New York Times Magazine,* February 22, 1970, p. 26.

42. This was a Navy proposal to lay down a huge grid of cables under 6,400 square miles of Wisconsin to serve as an invulnerable broadcasting antenna to the U.S. Polaris submarine fleet. McClintok and his collaborators argued that the system would not only be a massive insult to the Wisconsin environment but that it also would be ineffective. (See Michael McClintok, Paul Rissman, and Alwyn Scott, "Talking to Ourselves," *Environment,* September 1971, p. 16.) Public opposition in Wisconsin eventually reached the point that the Navy decided to try to find a home for its antenna in another state. Subsequently the project was abandoned altogether.

CHAPTER 13

Stopping Sentinel

Sentinel is, among other things, an anti-ballistic-missile shield that everyone agrees could not stop a concentrated missile attack, a strictly defensive system that its critics consider more belligerent than our current policy of keeping enough offensive missiles to make any attack suicidal, a five- or ten-billion-dollar "thin" shield against the Chinese (who have no missiles) which many people think will grow into a fifty- or hundred-billion-dollar "thick" shield against the Russians (who have too many to be affected by a thick shield), a boondoggle according to Dwight Eisenhower, a sensible compromise according to Robert McNamara, a "pile of junk" according to the prevailing view among scientists, and a functioning national program by act of Congress. . . .

At public meetings, the Army has shown Lake County [Illinois] citizens color slides of the computer-operated nuclear-defense system designed to protect them and their loved ones from what are commonly referred to as "primitive Chinese missiles" (conjuring up visions of thousands of Chinese peasants laboriously carting the mud of the Yangtze to crude molds, creating out of the baked earth something that roughly resembles an intercontinental ballistic missile, straining together to pull it back on some enormous catapult, and launching it seven thousand miles over the Pole in an attempt to obliterate Chicago). But the same meetings have almost always included a scientist from the Argonne National Laboratory, a center for non-military nuclear research just west of Chicago; explaining that he is

178

speaking not as an official representative of the laboratory but as a private citizen who happens to be a nuclear physicist, he reminds everyone that an unauthorized explosion is possible, even though extremely unlikely, and that such an explosion would destroy from a hundred and fifty thousand to two million citizens, "depending on which way the wind is blowing."[1]

—Calvin Trillin,
in *The New Yorker*

The Sentinel antiballistic missile (ABM) system was the Johnson administration's response to the threat of a new election year "missile gap," an application by the Republicans of the tactic that had helped elect John F. Kennedy in 1960.[2] The Sentinel system accomplished its prime political objectives: it successfully mollified the military establishment and blunted Republican criticism. But despite bipartisan Congressional support, Sentinel fell victim soon after the election to the powerful but largely unforeseen opposition of irate suburbanites across the country who wanted no nuclear bombs in their backyards. This chapter tells the story of the scientists who informed and helped organize the opposition to the Sentinel ABM system.

Defending the Cities

Had I known then what would occur, I never would have let it happen. I would have said [that placing ABM sites further away from] major cities would have been reasonable. I just didn't foresee the outcry of the cities.[3]

—Dr. Daniel Fink,
Deputy Director of Defense
Research and Engineering

The fifteen Sentinel ABM bases initially envisioned might have come into being if it were not for the impolitic enthusiasm of Director of Defense Research and Engineering Dr. John Foster and his deputy Dr. Daniel Fink, who decided to place several of these ABM bases in major American metropolitan areas. The threefold mission of the Sentinel ABM system announced by Secretary of Defense McNamara in 1967 was (1) to provide a thin "area defense" of the entire United States against missile attack by China, assuming that China would soon develop the capability of launching nuclear missiles against the United States; (2) to provide protection against a nuclear missile "accidentally" launched by the Soviet Union; and (3) to provide—"as a concurrent benefit"—a very limited defense of U.S. land-based Minuteman intercontinental nuclear missiles against Soviet attack.[4] None of these objectives tied missile sites to large cities, since the Sentinel ABM system depended primarily on the Spartan missile, with a range of some 400 miles. Indeed, the only rationale for placing ABM sites near cities was the possibility thereby provided of enlarging the system into a massive defense of population centers against Soviet missiles—a mission which Secretary McNamara had explicitly rejected as not feasible at any price.[5] McNamara feared that any attempt to defend our cities against a major missile attack would only inspire the Soviet Union to further escalate the arms race. But McNamara's preoccupation with Vietnam and his transfer out of the Defense Department soon after the decision to deploy Sentinel left effective control of ABM deployment in the hands of Deputy Secretary of Defense Paul Nitze and research and engineering chief Foster. Both these men favored keeping open the option for a large ABM system,[6] as did the Joint Chiefs of Staff and leading Congressional "Hawks".[7] Consequently, when the army announced, on November 15, 1967, the first ten areas to be surveyed for ABM sites, it transpired that eight were near major cities: Boston, Chicago, Dallas, Detroit, Honolulu, New York City, Salt Lake City, and Seattle.[8]

The fact that the proposed Seattle ABM site was actually within the city limits seemed especially puzzling to Newell Mack, a graduate student of biophysics at the University of Washington in Seattle. Mack had been interested in strategic weapons issues for several years and had discussed the arguments for and against missile defense with the experts. He now wrote to one of them, Hans Bethe:

Newspaper reports say Sentinel sites may be placed near cities and these sites are to be protected by Sprint missiles. In Seattle, at least, the proposed site of the Sentinel base with accompanying Sprint missiles is five miles from the heart of the city. I don't know whether Sprints are to be placed so close to other cities "tentatively chosen as possible locations" for Sentinel bases. . . . [If so,] the "thin" defense begins to look like a destabilizing "thick" defense.[9]

The short-range, quick-accelerating Sprint had originally been designed for urban defense as part of the massive Soviet-oriented Nike-X ABM system, which was proposed in 1963 but never deployed. In the Sentinel system, the Sprint was relegated to the more limited task of defending Minuteman missile fields and the large and vulnerable ABM radars.[10] The placing of the ABM radars and Sprints in major cities appeared to Mack and other observers as a regression to the old Nike-X population-defense concept, and, as such, an escalation of the arms race that would be likely to provoke a Soviet response.[11]

Mack was able to learn the exact sites being considered for the Sentinel bases in a number of other metropolitan areas by writing to the local newspapers and city officials. By early summer 1968 he was able to inform Representative Brock Adams (D.-Wash.) of Seattle that in at least seven of the first ten announced Sentinel locations, the proposed sites were indeed very close to population centers. Representative Adams inserted Mack's report into the *Congressional Record*, along with reports on other aspects of antiballistic missiles by several of Mack's colleagues at the University of Washington.[12]

Spreading the Alarm

The issue of ballistic missile defense was hardly new in 1968. Although the major ABM systems proposed after *Sputnik*—Nike-Zeus and Nike X—were opposed successfully by scientific advisors and others within the executive branch, enough of the controversy had spilled over into Congress and the press (especially journals like the *Bulletin of the Atomic Scientists*) for interested outsiders to follow the main arguments. Thus by the late 1960s there was widespread agreement among politically liberal and moderate scientists on the need for general arms-limitation agreements on offensive and defensive weapons, including ABMs. Indeed, a number of American scientists at the international "Pugwash" meetings[13] on arms control found themselves explaining to their Soviet counterparts why the rudimentary Galosh ABM system around Moscow was not perceived in the United States as the Soviets professed to see it, namely as a purely defensive system. Instead, by threatening to diminish the population-destruction capability of the American offensive missiles ("threatening the deterrent" is the jargon), the Moscow installation, numbering less than 100 interceptors, had given the Pentagon an excuse to develop thousands of multiple independently targetable reentry vehicles (MIRVs) for Minuteman III and Poseidon missiles.[14]

By early 1968, a small number of scientists—ranging from prominent government advisors to graduate students like Newell Mack—had begun to present the case against Sentinel to their professional colleagues and to the public. The Council for a Livable World, a scientists' political fund-raising group founded in 1962 by physicist and author Leo Szilard, organized anti-ABM symposiums for Senators and their aides; and the Federation of American Scientists, a public-education and lobbying organization founded in 1946, adopted position papers against ABM. Probably the most influential document in convincing scientists to oppose the ABM, however, was an article on the subject by Hans Bethe and Richard Garwin published in the March 1968 *Scientific American.*[15]

Bethe, a Nobel Prize-winning Cornell physicist, had been advising the government on strategic weapons since World War II, during which he was a leading figure in developing the atomic bomb. He had long opposed ABM deployment in his advisory capacity. When he saw the pressures for deployment increasing within the Johnson administration, he decided to try to prepare scientists outside government for the public debate which was to come. In June 1967 he delivered a talk at the University of Wisconsin in which he pointed out the great technical difficulty of effective missile defense.[16] After the Johnson administration's decision that fall to deploy Sentinel, Bethe reworked his talk and successfully sought permission from the Defense Department to include previously classified material. Bethe's revised talk was presented in a symposium at the annual meeting of the American Association for the Advancement of Science in December 1967. Richard Garwin, the IBM physicist who later played an important role in the SST debate, presented additional technical and strategic arguments against ABM deployment. Gerard Piel, publisher of *Scientific American*, happened to be present, and he urged the two scientists to write up their talks for publication in his magazine. The Bethe-Garwin article, along with the writings on ABM by David Inglis, Ralph Lapp, Leonard Rodberg, Jeremy Stone, and others,[17] provided essential background information for the scientists and laymen who organized to oppose the Sentinel sites in their own localities.

Seattle

In Seattle, the first inkling of the location of the Sentinel site came in April 1967 when the Army halted proceedings transferring title to Fort Lawton to the city. Seattle had long been planning to turn the old unused Army base, which is located in a heavily populated part of the city, into a civic park. Thus the Army's first opponents over the issue of Sentinel sites were the mayor and environmentally concerned Seattle citizens.

Scientists at the University of Washington decided to become involved when the Army's purpose in retaining Fort Lawton became clear in November 1967, a few weeks after Defense Secretary McNamara had announced the Sentinel deployment decision. In July 1967, Newell Mack had invited Hans Bethe to talk on ABM before the Graduate Conflict Studies Group, a seminar led by physics professor Gregory Dash. Bethe's talk generated considerable interest, and the group afterward discussed with him the possibility of assembling an anthology of pro- and anti-ABM literature. Bethe agreed to help, in the expectation that "next year [i.e., 1968] may well be the year of decision on U.S. deployment of an ABM system. It is essential that the public be informed and develop some opinion on it."[18] The Johnson administration's deployment decision came even earlier than Bethe had expected. Instead of working on the anthology, the University of Washington group—by then organized as the ABM committee of the Seattle branch of the Federation of American Scientists—bent their efforts toward briefing the mayor and other officials and assisting local citizens' groups fighting against the use of Fort Lawton as an ABM base.

Besides arguing against the Sentinel system as a whole on the grounds that it could be easily circumvented, penetrated, or saturated, the Seattle scientists particularly emphasized that the Spartan's long range in any case permitted the Sentinel base to be located some distance away from Seattle.[19] They also pointed out that the urban siting would make Seattle a particularly choice target—a "megaton magnet," to use Ralph Lapp's phrase—and in addition would needlessly expose a large population to the danger of an accidental nuclear explosion. The Army's local public relations people disputed these arguments, but the scientists stood their ground. They were reassured of the soundness of their position after Senator Henry Jackson (D.-Wash.) arranged a classified briefing for Edward Stern, a University of Washington physicist who happened to have security clearance: Stern reported back that, while he could not give details, the scientists' arguments were right.

By autumn 1968, a large coalition of Seattle citizens' groups, with members as diverse as the local Audubon Society and Junior Chamber of Commerce, had organized to oppose the Fort Lawton Sentinel site. Eventually, even Senator Jackson, who for years had been one of the staunchest supporters of ABM, was moved by the citizen pressure in Seattle to concede that perhaps another site could be found. With Jackson's assistance, the coalition persuaded the Army in December 1968 to shift its proposed missile site to a fashionable Seattle residential section, Bainbridge Island in Puget Sound. There, however, it again ran into determined opposition from local residents—fortuitously including another Congressional "Hawk," Representative Thomas Pelly (R.-Wash.)—who urged the Army to move the site someplace else.[20]

The Argonne Scientists

> We felt like a mouse crawling up an elephant's leg with thoughts of assaulting the elephant. Well, maybe we didn't succeed in that, but we made the elephant twitch a little.[21]
>
> —Dr. Stanley Ruby,
> physicist at Argonne National Laboratory
> and president of the Chicago Chapter
> of the Federation of American Scientists

In Seattle, the scientists were an essential auxilliary force in the citizens' coalition that opposed the local Sentinel ABM sites, but the main locus of the anti-ABM campaign was in the mayor's office. In Chicago the situation was reversed. There a few scientists at the AEC's Argonne National Laboratory, southwest of Chicago, were from the beginning at the center of the fight against the ABM.

In late October 1968, John Erskine, a physicist at the Argonne National Laboratory, was startled to read in his local community newspaper that

the Chicago base of the Sentinel Missle [sic] Air Defense System will be located either on a portion of the Healy farm land . . . or west of Westchester. . . . Both Spartan and Sprint missiles would be kept at the Chicago site, Col. H. G. Fuller, executive officer of the North Central Division, Army Corps of Engineers, Chicago, said. . . . Fuller added that residents surrounding the site would have no problem with excessive noise. . . . "These are not the type of missile with engines that can be warmed up," he said.[22]

Erskine was a member of a small group of Argonne scientists, mostly nuclear physicists, who had for several years been meeting regularly over lunch for discussions about the arms race and arms-control problems under the leadership of David R. Inglis, a former chairman of the Federation of American Scientists. The implications of ABM deployment had been discussed and dissected by this group for some time, so that when Erskine reported his news to them the more active members immediately understood its importance. Although some of the scientists were not opposed to deployment of a "thin" ABM defense, none of them believed that there was any justification for placing missiles with nuclear warheads within metropolitan areas. They all agreed that the citizens of Chicago should be given an opportunity to decide whether they wanted such neighbors.

Before taking the issue to the public, the Argonne scientists worked for two weeks to prepare themselves. They studied the available literature on ABM, including Congressional hearings; they even telephoned queries directly to the Sentinel System Command Base in Huntsville, Alabama. Finally, Erskine and Inglis contacted friends in the press and local television stations. And on November 15, 1968, the citizens of Chicago awoke to two-inch headlines warning of "A-Missile Sites in Western Suburbs."[23] Because the story had originated with the Argonne scientists rather than an Army press release, it was not written so as to allay fears about living in close proximity to hydrogen bombs.

Citizen protests began immediately. When Erskine returned home that evening, his telephone would not stop ringing: "People kept asking 'Hey, what can we do to help.' "[24]

In the next few weeks, the Argonne scientists talked with newspaper editors, Congressmen, mayors, and village officials. More than a dozen television interviews helped them tell their story to the Chicago area. They prepared a position paper and an information packet, and they helped to arrange public meetings to discuss the Sentinel system. An example will indicate how their influence pervaded the debate: when the Army organized a briefing session for local Congressmen and government officials, the questions of safety that had been raised by the Argonne scientists prompted one member of the audience to ask how far from Chicago the missile site could be placed without reducing its effectiveness. The speaker, Colonel William Wray, chief of site operations for the Sentinel System Command, refused to answer the question "for security reasons."[25] The Argonne scientists were there, however, and pointed out to the press and television media afterward that the answer could be deduced from the well-advertised range of the Spartan missile.

"ANYWHERE EXCEPT NEAR US"

The residents of Westchester, one of the suburban communities west of Chicago, whose town dump had been chosen as a possible Sentinel site, were not enthusiastic.[26] "We'd rather have the dump," explained one housewife who was circulating a petition:

We all realize that the dump is a temporary thing. After 20 years or so they will turn it into a golf course. But the missile site is more permanent and it can't do any good to our property values. Besides its unattractiveness, there is also the danger. They say that they haven't had an accident in 20 years, but if they have the first one here, we won't be around to tell about it.[27]

The Westchester Village Board scheduled a special meeting on the issue on December 3, 1968, and invited the Army to send a representative. Army officials declined, however, claiming that their representatives could not attend because information about the sites was classified.[28] This stategy proved to be unsuccessful. At the meeting the local representative, Harold R. Collier (R.-Ill.), criticized the Army for making it difficult to present an evenhanded informational session. He added that he personally strongly opposed the Westchester site because of the danger of an accidental explosion, and he informed the citizens of Westchester that Congress had been "assured the system would be placed in sparsely settled areas."[29] Two scientists from Argonne, John Erskine and John Schiffer, also spoke at this meeting. At the end of the meeting the audience was convinced—all but about 25 of the nearly 400 people in attendance raised their hands to indicate opposition to the Sentinel site. The village board responded by unanimously adopting a resolution to the same effect.[30]

That same day, after hearing Argonne physicist George Stanford describe the likely effects of the accidental explosion of a Spartan warhead,[31] the Executive Committee of the DuPage County Board went on record opposing ABM sites anywhere in the Chicago area.[32] Three days later, the York Woods Community Association passed an equally strong resolution after hearing from John Erskine and Roy Ringo (yet another Argonne physicist). Army officials had once again declined to appear.

Thus, largely as a result of the efforts of the Concerned Argonne Scientists, the ABM was "invited out" of the western Chicago suburbs. On December 12, the Army responded by announcing that it had decided to locate the Chicago-area ABM site in an abandoned Nike-Ajax base near Libertyville, a suburb north of Chicago.

LIBERTYVILLE, ILLINOIS

> *The Army is not here to debate the government's position to deploy the Sentinel Ballistic Missile [sic] in the Libertyville area. We cannot discuss the political aspects of the issue. We have been told what to do.*
>
> *We are hopefully here to develop a meaningful dialogue on the Sentinel missile.*[33]
>
> —Colonel R. J. Bennett,
> Army information officer

Libertyville is a more conservative community than the towns west of Chicago where the Argonne scientists had hitherto campaigned. The Libertyville area residents reacted calmly to the news that Spartan missiles with their multi-megaton warheads were to be their new neighbors. Libertyville Mayor Charles Brown expressed the general reaction:

The almost miraculous technology of our world today has far surpassed our meagre ability to comprehend. Under these circumstances, it would certainly seem more prudent to place our confidence and security in the hands of those whose lives are dedicated to the profession of defending and protecting our lives, our loved ones, and our properties than to try to accumulate sufficient knowledge to make an independent decision.[34]

But Clarence Pontius, supervisor of Vernon Township, a thirty-six-square-mile area which includes the missile site, said he wanted to know more about the project:

I've heard some of the Argonne scientists describe the dangers on television, but it seems to me there's insufficient information. They made flat statements and didn't back them up. I want to hear more from the Army Engineers.[35]

This time the Army, anxious not to repeat its debacle in the western Chicago suburbs, dispatched its top team. On December 19, 1968, Lieutenant General Alfred Starbird, manager of the Sentinel system for the Army, and John S. Foster, Jr., Director of Defense Research and Engineering, flew out from the nation's capital to present the Defense Department's case in a briefing for officials of the northern Chicago area. The briefing was also open to the public. Both of the Pentagon representatives insisted that the Sentinel site had to be as close as possible to Chicago in order to protect the city from the threat of a Chinese Communist attack; and Foster even admitted that he expected the Sentinel ABM might "thicken" into a defense against Soviet missiles depending "on the nature of emerging technology."[36] Responding to the citizens' concerns, General Starbird insisted: "There cannot be an accidental nuclear explosion."[37]

Meanwhile, in the audience, John Erskine and other Argonne scientists quietly handed out leaflets containing a map of the sixty-square-mile area that would be flattened and incinerated if one of the warheads nevertheless did explode. The leaflet also pointed out that, if the winds were right, fallout would kill much of the population of Chicago.

When invited to confront the Argonne scientists, Starbird and Foster replied that they had to leave immediately for Washington. The Argonne scientists then spoke to the remaining townspeople and newsmen. John Erskine pointed out that "the Army let the cat out of the bag" by admitting that Sentinel had become a city defense.[38] George Stanford labeled the Army's claim that a nuclear accident is impossible "a ridiculous statement. . . . They have circumvented a lot of possibilities, but they still have the human and mechanical components to consider."[39] The Argonne scientists then quoted from the government's official nuclear weapons handbook:

Nuclear weapons are designed with great care to explode only when deliberately armed and fired. Nevertheless, there is always a possibility that, as a result of accidental circumstances, an explosion will take place inadvertently. Although all conceivable precautions are taken to prevent them, such accidents might occur in areas where the weapons are assembled and stored, during the course of loading and transportation on the ground, or when actually in the delivery vehicle, e.g., an airplane or a missile.[40]

The scientists emphasized that ABM warheads would be particularly difficult to safeguard against accident because they must remain ready to be launched and exploded on a moment's warning: a hair trigger cannot simultaneously be a stiff trigger.

A few days after the Foster-Starbird briefing, local newspapers were quoting the previously "unconvinced" Vernon Township supervisor, Clarence Pontius, repeating the same arguments against locating an ABM site in the Libertyville area used by the Argonne scientists.[41]

With one village board after another voting to oppose the Libertyville ABM site, the army finally decided to try to counter the remarkable effectiveness of the Concerned Argonne Scientists by fielding a public information team of its own. The Army team, while it lasted, ordinarily consisted of

two full colonels (one of whom introduces the other), a lieutenant-colonel working the slide projector, and a civilian public-relations man with a pipe, a Sentinel tie clasp, and an elaborate tape recorder.[42]

Both the scientists and the Army spokesmen toured Lake County, Illinois, "like old prizefighters staging exhibitions"[43]—but after about a month the Army gave it up. The more the citizens heard, the more they organized to oppose ABM. In mid-January, one of these anti-ABM groups filed suit to stop construction on the Libertyville site pending judicial and Congressional review. A federal district judge, after agreeing to assume jurisdiction, warned the Army not to start construction until he rendered his decision; and on March 3 he denied a government motion to dismiss the suit.[44] Around the same time in March, coinciding with protests at MIT and other leading universities against the military's misuse of science, faculty members and students at Northwestern and other Chicago-area universities finally began to express opposition to the Sentinel ABM system.[45] Meanwhile, citizen protests in other metropolitan areas being considered for ABM sites also began to receive national attention.

Reading, Massachusetts

The people against the site are playing a game of Russian roulette with the survival of this country. . . .Scientists at M.I.T. have apparently accepted the Boston site, which is closer to the central city area than the Vernon Hills [Illinois] site. There has been no disapproval from M.I.T.[46]

—Representative Roman Pucinski (D.-Ill.)

In Detroit, two physicists from local campuses conducted an anti-ABM campaign much like that of the Argonne group, although on a smaller scale.[47] But politically active scientists in the Boston area—home of Harvard, MIT, and a dozen other academic centers—were too busy commuting to Washington to concern themselves with the ABM site construction that had already begun north of Boston. There the local citizens led the opposition from the beginning.

One community went so far as to appropriate $2,500 for "appraisal, engineering, legal and other expenses" to conduct a study of the implications of the proposed ABM site "for the purposes of protecting the interests of the town."[48]

The Boston-area Sentinel opposition culminated in a confrontation in Reading, Massachusetts, the site of one of the two Massachusetts ABM installations. The New England Citizens Committee on ABM had responded to the Army's announced briefing by drumming up a large crowd and recruiting a distinguished anti-ABM panel, including ex-Presidential science advisor Jerome Wiesner, former high-ranking Defense Department weapons analyst George Rathjens, and Kennedy aide Richard N. Goodwin. Patrick J. Friel, former Deputy Assistant Secretary of the Army for Research and Development, attended the meeting. The next morning he wrote to John Foster:

I was very impressed with the fact that the audience was extremely well informed and would not accept weak answers on either the technical or policy aspects of the system. It is fairly clear to me that a substantial fraction of the people present (over 2,000) fully intend to prosecute the issue further with their congressmen and senators. . . . If this is the typical reaction throughout the country, and if the information exchange continues to be as inadequate as last night's presentation in Reading, it seems to me that there is a very good chance that the Congress would have to act to cancel the system.[49]

Senator Edward Kennedy (D.-Mass.) responded to the Reading meeting by firing off a long letter to incoming Defense Secretary Laird, calling on him to stop Sentinel deployment pending a complete review.[50] And Massachusetts Representative William H. Bates, the ranking Republican on the House Armed Services Committee, pressured the committee chairman, Mendel Rivers (D.-S.C.). Surprisingly, Rivers obliged by writing Laird suggesting that Laird's recent statements had indicated uncertainty about Sentinel. "If such is the case," Rivers wrote, "I think that before we proceed any further you should indicate to me what your probable course of action will be."[51]

Legislators in other parts of the country were also feeling the political heat; even Senate minority leader Everett M. Dirksen (R.-Ill.), who had been a stalwart defender of Sentinel, conceded that "perhaps the time has come to take a cooler and more deliberate look at this proposal."[52] Dirksen must have been getting a lot of mail on this issue: his junior Senatorial colleague Charles Percy (R.-Ill.), an early opponent of ABM, was receiving 750 to 1,000 letters a week on this issue from constituents—almost all of them expressing opposition to ABM. And Representative Chet Holifield (D.-Calif.), chairman of the powerful Joint Committee on Atomic Energy, also expressed doubts about the Sentinel system. (One of the proposed Sentinel sites was in a Los Angeles suburb only half a mile from Holifield's home.)

At the end of February 1969, Defense Secretary Laird announced that all work on the Sentinel system would be halted pending review. Two weeks later Sentinel was officially dead: on March 14, President Nixon announced that the ABM missile and radar sites would be removed from the cities to more remote locations, and he rechristened the system "Safeguard." The official rationale was changed along with the name: the primary purpose of Sentinel had been a light area defense against anticipated Chinese ICBMs (intercontinental ballistic missiles); the primary purpose of Safeguard was to be defense of the U.S. Minuteman ICBMs against a preemptive Soviet attack.

Postscript

Secretary of Defense McNamara had warned, in his 1967 speech announcing the decision to deploy the Sentinel ABM system, that "pressures will develop to expand it into a heavy Soviet-oriented ABM system."[53] That these pressures were successfully resisted was largely due to the rebellion of the suburbanites against bombs in their backyards. However, to the many scientists who opposed deployment of the Sentinel hardware in any location, this victory seemed rather hollow. They feared that by moving the missiles away from the cities, the Nixon administration would succeed in making an expensive and unnecessary ABM system politically practicable.

In retrospect, the campaign against Sentinel appears to have been much more significant in influencing ABM politics than was initially supposed. The potent citizen resistance to the Sentinel system made the whole subject of ABM a national issue and convinced both politicians and scientists that the ABM was an issue on which Congress should make an independent decision.

Let us briefly review the post-Sentinel ABM developments. Once the Nixon administration made the decision to move the ABM sites away from the cities, the focus of the debate turned to a question with which the technical experts were more comfortable: Would the proposed ABM system in fact provide a cost-effective missile defense? Defense Department officials, of course, argued uniformly in Congressional hearings that the answer to this question was affirmative, sometimes citing independent experts to buttress their arguments. But, as we have seen in Chapter 5, many of these experts were actually opposed to ABM, and in appearances before Congressional committees they followed Bethe and Garwin in outlining a variety of relatively inexpensive techniques that an attacker could use to penetrate the Safeguard system. ABM opponents also emphasized the vulnerability of ABM radars, the system's unprecedented complexity, the impossibility of testing it, and the limited nature of Safeguard's capabilities even if it should actually work as designed. ABM proponents meanwhile asserted that the continued Soviet offensive-missile deployment required some response and that any technical problems with the ABM could be overcome once a commitment to the system had been made. Thus there was less a debate than a standoff, with the ABM opponents concentrating on the system's technical limitations and the proponents concentrating on the potential Chinese or Soviet threat.[54]

In the face of strong opposition in the Senate, the marketing of the Safeguard system in 1969 was definitely softsell. In addition to moving the ABM sites out of the suburbs, the Nixon administration offered to finance the Safeguard system on the installment plan. Congress was asked only to authorize funding for two ABM sites to defend Minuteman ICBM bases in Montana and North Dakota. Authorization of additional sites was to be contingent on the demonstration to Congress that the ABM technology was indeed advanced enough to be effective. Defense Secretary Laird presented the argument as follows:

> To those who are concerned about whether the Safeguard system will work, I would say let us deploy phase 1 and find out. Only in this way can we be sure to uncover all of the operating problems that are bound to arise when a major weapons system is first deployed. Since it will take five years to deploy the first two sites, we will have ample time to find the solutions through our continuing R&D [research and development] effort to any operational problem that may arise. And only then will we be in a position to move forward promptly, and with confidence, in the event the threat develops to a point where deployment of the entire system becomes necessary.[55]

With this assurance and partially persuaded by the administration that the Safeguard system was an essential "bargaining chip" in the strategic-arms limitation talks (SALT) with the Soviet Union the Senate in 1969, as a result of a tie-breaking vote cast by Vice President Agnew, decided to let the deployment proceed.[56]

The following year, however, the Nixon administration was back asking for funds to begin ABM deployment on a third ABM site in Missouri and to acquire land and do preliminary work on another five sites. There was widespread anger in the Senate at the administration's abandonment of its commitment of the year before, and even the hawkish Senate Armed Services Committee began to find some merit in the arguments of technical experts who appeared before it opposing further deployment. These witnesses pointed out that none of their technical criticisms of the Safeguard system design had been answered in the intervening year.[57] They also pointed out that the Chinese had still not tested a missile which could deliver a nuclear warhead to the United States,[58] whereas the imminence of such a Chinese capability had been the primary justification put forward for immediate deployment by Secretary McNamara three years before.

Thus in June 1970 the Senate Armed Services Committee, while approving ABM sites to defend two additional Minuteman bases against possible Soviet attack, refused to approve another four sites whose primary purpose would have been to defend against a Chinese attack.[59] The approval of even the two additional sites barely passed the Senate after a White House aide showed wavering Senators a telegram from the chief U.S. negotiator at the SALT talks claiming that ABM expansion was essential to the success of the talks.[60]

In 1971 the Nixon administration asked Congress for the option to build an ABM site to defend Washington, D.C., instead of one of the four sites defending Minutemen bases. But the Senate Armed Services Committee refused even this limited request—giving as its reason the fact that schedules in the rest of the program had slipped by almost an entire year and that the army was not yet ready to proceed with additional bases. Finally, in May 1972, the United States signed the SALT-I agreement with the Soviet Union limiting ABM deployment in each nation to a total of 200 ABM missiles deployed at two sites—one to be located near the capital of each nation (the Soviets had already deployed the primitive Galosh ABM system around Moscow) and one other site (corresponding in the United States to one of the sites defending Minuteman bases). In Congressional testimony Defense Secretary Laird indicated that he had gone along with this agreement because he had concluded, after three-and-a-half years of trying, that the administration would not succeed in getting Congress to authorize the full national Safeguard deployment.[61]

The battle over the ABM sites in the suburbs had served effectively to raise the entire issue of missile defense to a level of visibility where Congress was able to act for once as an equal branch of government in setting national defense policy. The outcome was quite different from what it might have been had the decisions made inside the executive branch been final.

Stopping Sentinel: An Analysis

The activities of scientists all across the country were important in stopping Sentinel. In fact, the geographical coverage of the opposition was perhaps its most important source of strength, particularly in its impact on Congress. But the greatest credit for Sentinel's demise must go to the indefatigable scientists from Argonne National Laboratory. In fact, a special Defense Department analysis of national editorial reaction found that in late 1968 newspapers which had previously supported the Sentinel program began opposing it, "when the major protest movement started last mid-November in Chicago, led by a group of nuclear physicists."[62]

What accounts for the Argonne group's success? Dedication, certainly. George Stanford estimates that he personally participated as a speaker or debator on at least thirty occasions and that three others—David Inglis, John Erskine, and Stan Ruby—were about equally active. In all, ten Argonne people made one or more speeches against the ABM. This activity was not without personal sacrifice: several used vacation time for their anti-ABM activities and spent hundreds of dollars each for transportation and telephone bills.

Another essential element in the Argonne scientists' effectiveness was their excellent relations with the press. They were the first to reach the key local media with the news of the planned missile sites in the Chicago suburbs. They maintained their good press relations by doing their homework, so that they could not be caught in careless errors, and by preparing clear and well-written statements of their views for public distribution.

Finally, the most important reason for the success of the Sentinel opposition lies in the fact that the arguments against "bombs in the backyard" struck such a responsive chord with the public. Ironically, however, the fact that this issue was the key to obtaining public attention for the ABM controversy has been a source of some disillusionment to anti-ABM scientists. Most of them considered the dangers inherent in an uncontrolled arms race to be much more serious than the danger of an accidental nuclear explosion in the suburbs. But the public has been largely silent during the quarter-century since the destruction of Hiroshima and Nagasaki while the military in both East and West has stockpiled enough nuclear weapons to destroy civilization in the next total war. Only when the nuclear arms race threatened to become a concrete local reality were suburbanites prodded into action.

Were the Argonne scientists irresponsible in using the possibility of an accidental explosion to "wake people up," as David Inglis put it?[63] It is true that the possibility of an ABM warhead exploding accidentally or as a result of human error or sabotage is remote. But the Argonne scientists asserted that the possibility existed—and that it indeed might well be as great as the possibility of a missile attack on Chicago. They felt that even a small chance of a great catastrophe should not be taken lightly, especially when they could find no counterbalancing benefits, and they saw to it that the citizens who were asked to bear such a risk were informed and had a voice in the decision.

NOTES

1. Calvin Trillin, "U.S. Journal: Lake County, Ill.," *New Yorker*, February 15, 1969, pp. 100-106.

2. For more on the history of antiballistic missile programs, see Chapter 5.

3. Quoted in Anne Hessing Cahn, *Eggheads and Warheads: Scientists and the ABM* (Cambridge: Center for International Studies, MIT, 1971), p. 188. We found Dr. Cahn's dissertation to be very helpful in researching this chapter. An abbreviated version appears in *Science and Public Affairs*, ed., Albert H. Teich (Cambridge, Mass.: MIT Press, 1973).

4. Speech to United Press International editors and publishers, San Francisco, September 18, 1967, quoted in the *New York Times*, September 19, 1967, p. 18. See also Henry L. Trewhitt, *McNamara* (New York: Harper & Row, 1971), esp. pp. 123-132.

5. *New York Times*, September 19, 1967, p. 18.

6. See Cahn, *Eggheads and Warheads*, particularly the statement of former Deputy Assistant Secretary of Defense Martin H. Halperin quoted on pp. 187-88.

7. For example, in the secret Senate ABM debate of October 2, 1968, Senator Richard Russell (D.-Ga.), chairman of the Senate Armed Services Committee, said: "It was my own view that [Sentinel] was a system that could, of course, be used against any Chinese threat, but I considered it to be—and I want to be frank . . . and not deceive anyone—the foundation stone of a missile defense system . . . to protect the people of this country against a Soviet missile atomic attack. . . . If we have to start over again with another Adam

and Eve, then I want them to be Americans and not Russians, and I want them on this continent and not in Europe." [*Congressional Record* 114 (1968): 29170-29175.]

8. The other two sites were Albany, Georgia, and the Grand Forks, North Dakota, Minuteman base. Five additional sites were announced later: Los Angeles; San Francisco; Cheyenne, Wyoming; Great Falls, Montana; and Sedalia, Missouri.

9. Letter from Newell Mack to Hans Bethe, December 19, 1967.

10. See the testimony of Dr. Daniel Fink in U.S. Congress, Senate, Committee on Foreign Relations, *Strategic and Foreign Policy Implications of ABM Systems*, 91st Cong., 1st sess., March 6, 1969, pp. 22-30.

11. After the Nixon administration had decided to substitute Safeguard for Sentinel, Secretary of Defense Melvin Laird attacked Sentinel as follows: "In the discussions which preceded the authorization of this Sentinel system by the Congress, it had been publicly stated that the system did not have a capability of defending our cities against a heavy attack of the kind the Soviets could launch. . . . It was obvious, however, . . . that the Sentinel system was ambiguous, at best. It was interpreted by some as the beginning of a 'thick' defense of our cities against Soviet attack. In fact, it could have been used for precisely that purpose. It could also have been construed as a system designed to protect our cities from surviving Soviet missiles after a surprise attack by the United States. Our review, therefore, convinced us that the original Sentinel was potentially provocative. As such, it appeared to us to be a step toward, rather than away from, an escalation of the arms race." *Strategic and Foreign Policy Implications of ABM Systems*, March 21, 1969, p. 168-69.

12. *Congressional Record* 114 (1968): 20699-20704. The other reports were prepared by J. Gregory Dash, Philip A. Ekstrom, Diane M. Hartzell, and Edward A. Stern.

13. For more on Pugwash, see Chapter 11, note 11.

14. Early generations of intercontinental ballistic missiles (ICBMs) and submarine-launched ballistic missiles (SLBMs) carried single-bomb warheads. The most advanced generation of American ICBMs (e.g., Minuteman III) and SLBMs (e.g., Poseidon) are equipped with multiple independently-targeted re-entry vehicles (MIRV) warheads, in which from three to a dozen nuclear weapons can be individually dispatched to pre-selected targets. The Soviet Union first tested MIRVs in 1973.

15. Richard L. Garwin and Hans A. Bethe, "Anti-Ballistic Missile Systems," *Scientific American*, March 1968, pp. 21-31. Our description of the events leading up to this article is from Cahn, *Eggheads and Warheads*, pp. 90-91.

16. This talk was the first lecture in an annual series on the impact of science on society established in memory of Professor Julian E. Mack of the University of Wisconsin Physics Department (Newell Mack's father).

17. There were numerous articles by these and others in *Bulletin of the Atomic Scientists* and elsewhere, including books by Jeremy Stone, *Containing the Arms Race* (Cambridge: MIT Press, 1966), and Ralph Lapp, *The Weapons Culture* (New York: Norton, 1968; Baltimore: Penguin, 1969). Anti-ABM anthologies include Eugene Rabinowitch and Ruth Adams, eds., *Debate the Antiballistic Missile* (Chicago: Bulletin of Atomic Scientists, 1967), and Abram Chayes and Jerome B. Wiesner, eds., *ABM* (New York: Harper & Row, 1969). Pro-ABM anthologies include Johan J. Holst and William Schneider, Jr., eds., *Why ABM? Policy Issues in the Missile Defense Controversy* (Elmsford, N.Y.: Pergamon, 1969), and William R. Kintner, ed., *Why the ABM Makes Sense* (New York: Hawthorne, 1969).

18. Letter from Hans Bethe to Newell Mack, August 21, 1967. Much of the information in this and the following paragraph comes from interviews with Mack. See also Cahn, *Eggheads and Warheads*, pp. 42, 50, 242.

19. *Congressional Record* 114 (1968): 20701f.

20. *Congressional Quarterly Weekly Report*, December 13, 1968, pp. 3275-77. *New York Times*, February 9, 1969, p. 1. Thomas A. Halstead, "Lobbying Against the ABM," *Bulletin of the Atomic Scientists*, April 1971, p. 23.

21. Quoted in "ABM Furor: Four quiet men at Argonne Lab started it all," *Chicago Sun-Times*, August 7, 1969, p. 38. Argonne National Laboratory is an AEC installation for research on nuclear reactors, basic research in physics, chemistry, and biology and other non-military subjects.

22. (La Grange, Ill.) *Suburban Life*, October 26, 1968. Quoted in Cahn, *Eggheads and Warheads*, p. 71. It was later indicated that the Chicago site would house only Spartans, not Sprints.

23. *Chicago Sun-Times*, November 15, 1968, p. 1.

24. Quoted in *Chicago Daily News,* February 7, 1969, p. 6.

25. *Chicago American*, November 20, 1968. Quoted in Cahn, *Eggheads and Warheads*, p. 73.

26. One local headline read: " 'Anywhere Except Near Us': Westchester Fights A-Base." *Chicago Daily News*, December 4, 1968.

27. Quoted in *Suburban Life*, November 23, 1968, pp. 1-2.

28. *Chicago Daily News*, December 4, 1968.

29. Ibid.

30. *Chicago Sun-Times*, December 4, 1968, p. 16

31. The standard reference for such calculations is S. Glasstone, ed., *The Effects of Nuclear Weapons* (Washington, D.C.: Government Printing Office, 1962).

32. *Chicago Sun-Times*, December 6, 1968.

33. Quoted in the *Waukegan* (Ill.) *News-Sun*, January 10, 1969, pp. 1,4.

34. Trillin, "U.S. Journal," p. 106.

35. Quoted in the *Chicago Sun-Times*, December 13, 1968, p. 40.

36. *Chicago Sun-Times*, December 20, 1968, p. 6.

37. Quoted in the *Waukegan News-Sun*, December 20, 1968, p. 1; see also ibid., p. 20.

38. Quoted in the *Chicago Daily News*, December 19, 1969.

39. Quoted in the *Waukegan News-Sun*, December 20, 1968, p. 35.

40. Glasstone, *The Effects of Nuclear Weapons*, Appendix A, Section 1, p. 664.

41. *Waukegan News-Sun*, December 27, 1968, Regional Page.

42. Trillin, "U.S. Journal," p. 102.

43. Ibid.

44. *Chicago Sun-Times*, March 4, 1969.

45. *Chicago Tribune*, March 3, 1969, p. 22.

46. *Libertyville* (Ill.) *Independent-Register*, January 16, 1969.

47. The physicists were Drs. Alvin M. Saperstein of Wayne State University (Detroit) and William Hartman of Michigan State University. Saperstein alerted the local press, which responded with scare headlines: e.g., "Nuclear Missiles Slated for Suburbans, Peril Seen for Entire Metro Area," *Detroit Free Press*, December 1, 1968, p. 1. The two physicists were later joined in voicing opposition to the local ABM site by Detroit Congressman John Conyers (D.-Mich.).

48. Irving H. Shen, "The Anti-Anti-Missile Movement," *North Shore (*Mass.*) Weekend Supplement*, June 8, 1968. Quoted in Cahn, *Eggheads and Warheads*, p. 191.

49. Letter from Patrick J. Friel to John Foster, January 30, 1969. Quoted in Cahn, *Eggheads and Warheads*, pp. 192-193.

50. *New York Times*, February 2, 1969, p. 42; February 9, 1969, p. 1.

51. Quoted in the *Chicago Daily News*, February 7, 1969, p. 1.

52. *New York Times*, February 9, 1969, pp. 1, 64.

53. *New York Times*, September 19, 1967, p. 18.

54. See, e.g., Paul Doty, "Can Investigations Improve Scientific Advice? The case of ABM," *Minerva*, April 1972, Vol. 8, pp. 280-294. The primary forums in which these arguments were aired were the hearings before the Senate Foreign Relations Committee and Senate Armed Services Committee in 1969 and 1970.

55. Before the Senate Armed Services Committee, May 22, 1969; quoted by Senator Proxmire, *Congressional Record* 166 (1970): 25934.

56. Nathan Miller, "The Making of a Majority: Safeguard and the Senate," *Washington Monthly*, October 1969, p. 60.

57. Ibid. See also U.S. Congress, Senate, Armed Services Committee, *Authorization for Military Procurement, Research, and Development, Fiscal Year 1971*, 91st Cong., 2nd sess., March-June, 1970, Part 3.

58. Ibid.

59. *New York Times*, June 18, 1970, p. 1.

60. *New York Times*, August 14, 1970, p. 2.

61. *New York Times*, June 6, 1972, p. 1. See also U.S. Congress, House Committee on Appropriations, *Department of Defense Appropriations for 1973–Part 8*, 92nd Cong., 2nd sess., June 5, 1972, p. 379.

62. *Waukegan News-Sun* (AP), February 7, 1969, p. 1.

63. *Chicago Sun-Times*, August 7, 1969, p. 38.

CHAPTER 14

Public Interest Science in the University: The Stanford Workshops on Political and Social Issues

*Students looking at the Stanford cur-
riculum see little relation between the
courses being offered and the problems
of our society—urban blight and the
ghetto . . . outrageous influence of the
military . . . pollution and destruction
of the environment . . . .*

*And even where courses are directed
to the study of particular problems,
active engagement in possible solutions
is rarely considered.*

*We are a few students who feel that
the urgency of these problems warrants
a more active approach, and have
organized several workshops to study
issues of local and national concern
directly—specifically in order to consider what can be done about them.*

—from the first SWOPSI catalogue,
fall 1969

American universities possess on their faculties the nation's primary independent
reservoir of technical talent. It is natural therefore to look first to the

universities for leadership in public interest science—and most of the scientists in our case studies in preceding chapters have in fact been affiliated with universities.

The most potent combination that exists in the university—in public interest science, as in research—is the combination of the energy and enthusiasm of able graduate students with the knowledge and experience of faculty members. The success of some of the Stanford Workshops on Social and Political Issues (SWOPSI) illustrates the potential of this combination. The SWOPSI workshops were first organized by two graduate students and one undergraduate at Stanford University in fall 1969.[1] The subjects of these courses ranged from air pollution in the San Francisco Bay Area to international arms control and disarmament, and almost all of them were offered for full academic credit. Below we tell the stories of some of the more successful of these workshops.

The Logging Study

Allan Cox, a noted Stanford University professor of geophysics, lives in the rustic town of Sky Londa, California, located in the mountains of the Pacific Coast Range a few miles to the west of the Stanford campus. During 1968 he became concerned about both the increased logging in his area and the logging practices, which appeared to him to be unnecessarily destructive. By summer 1969 Cox and several of his neighbors were lobbying with the San Mateo County Board of Supervisors asking them to deny a logging permit for a proposed operation near Sky Londa. The county had previously passed ordinances to prohibit logging companies from leaving the forest floor littered with small dead timber and the streams choked with silt and debris. But attempts by the county to enforce these ordinances were fruitless. (Ultimately the courts ruled that the California Forest Practices Act of 1945, providing for self-regulation of the timber industry, completely preempted the field of logging legislation—despite the fact that this law made no provision for protection of the environment in urban areas.) It did not take long for Cox to conclude that better laws were required.

Dave Soper, a graduate student of physics at Stanford, agreed to join Cox in setting up a SWOPSI workshop on logging. Their goals were to identify the main social costs of logging in suburban areas, formulate a set of objectives for public policy on logging, analyze the effectiveness of current regulatory practices, and ultimately to generate recommendations for action. Brief descriptions of this and the nine other workshop-courses that were also organized during summer 1969 were combined to form the first SWOPSI catalogue, which was distributed at Stanford's fall 1969 registration. The student response was respectable if not overwhelming. Thirteen students registered for the logging workshop, of whom ten ultimately completed the course. The students came from a variety of academic backgrounds, but most had previously been interested in environmental issues.

During the course of the workshop its members interviewed logging company officials, forestry experts, a county tax assessor, planning commission staff members, and members of local conservation groups. In addition, most of the workshop went on a field trip to study a well-managed logging operation and also attended one or two county government hearings on logging.

The efforts of the logging workshop were devoted almost entirely toward preparation of a report, *Logging in Urban Counties.*[2] The students were assigned to write the various chapters: an overview of logging and man's environment, a history of the logging controversy in San Mateo and Marin counties, logging economics, and tax policy affecting logging. The entire group met for about two hours each week.

The workshop was unlike most academic courses in that its leader was not an authority on the subject being studied. Consequently, Professor Cox cast himself in the role of editor of the logging report rather than that of instructor. Most of the chapters went through at least one stage of detailed criticism and rewriting. The work at first showed a number of weaknesses: too little feeling for what constitutes a well-reasoned and well-documented argument, lack of experience in locating relevant government documents, and a tendency after interviewing a public official or a logger to write a personal emotional reaction rather than to give a factual account. Professor Cox did not hesitate to send the students back for another interview if the first try was unsatisfactory.

At the end of the three-month workshop, the students' work and the leaders' careful editing resulted in a well-written and thorough 100-page report. The technical background of the workshop leaders was reflected in a discussion of various models of forest management in the report (clear-cutting versus selective logging), as well as in the generally careful quantitative treatment of economic issues. The report was distributed to county and state officials, conservation groups, and the news media. Preparation of a short summary and a press release helped to increase the coverage given the report by the local Bay Area newspapers. As a result of this publicity, several hundred additional copies of the report were sold (at cost) during the next several months.

This concluded the workshop's official activities, and it was in fact the end of the involvement of most of the students. But the local logging situation was just beginning to be politically interesting. In February 1970, just after the SWOPSI logging report became available, the San Mateo County Board of Supervisors and the County Planning Commission met in an extraordinary joint session to consider the logging question. There was a large turnout of loggers and citizens groups, and good news media coverage. Cox and Soper made a formal presentation of their workshop's report.

The upshot was that the County Board of Supervisors decided to ask the local State Assemblyman and Senator to introduce a bill in the state legislature permitting a "local option" for counties to impose controls stricter than those of the State Forest Practices Act. The State Division of Forestry's District Rules

Committee met several times to enact special rules in an effort to placate the county without changing the state law. But the aroused county officials and conservationists were not so easily satisfied. They objected that the proposed rules lacked teeth for enforcement and that they ignored a crucial requirement— the appointment of individuals to the District Rules Committee who would represent the interests of the general public.

The focus of attention now shifted to Sacramento. During the spring and summer Cox, Soper, several housewives from Sky Londa, and a few officials of San Mateo and Marin counties joined in what Cox calls "low-grade lobbying" of the state legislature in favor of the "local option" bill. They had minimal help from established conservation organizations. The State Division of Forestry and the timber industry both opposed the proposed law, but 1970 was a year of great concern for the environment, and the fact that there was an election coming up in November helped the conservationists a great deal. The bill passed both houses of the legislature in September 1970. The local citizens group then worked hard through Republican contacts to get Governor Reagan to sign the bill—which he did.

Under the new law, San Mateo County officials immediately began the job of drafting county ordinances to regulate timber operations. In the early months of 1971 they held hearings to solicit input from loggers, land owners, conservation groups, and other interested parties. Informal shirt-sleeve sessions between all groups hammered out details. The final ordinance was passed in April 1971. Later that same month the timber company whose practices had most offended the conservationists announced that it was going out of business.[3]

The logging workshop had worked on a limited but significant problem, and its efforts had paid off. Professor Cox adds:

Our work on logging has had a strong impact on my own life and on that of several students—new career directions, fresh motivations, even new (and deep) friendships. Not very important on the scale of national problems, but important on the scale of individual lives.[4]

Air Pollution

Another one of the first ten SWOPSI workshops ambitiously tackled the problem of air pollution in the six-county San Francisco Bay Area. Some sixty undergraduates, twelve graduate students (including eight law students), a faculty member, a medical doctor, and a housewife participated in the workshop, which was led by Edward Groth III, a graduate student of biology at Stanford. Unlike Allen Cox, Groth was already an expert on the subject of his workshop, since the study of air pollution was a major part of his doctoral research. He consequently took a rather active role in the direction of the workshop, beginning with several introductory lectures on the nature of pollution problems. As the workshop progressed, however, Groth's role, like Cox's, became increasingly that of editor-in-chief, supervising the work of eight contributing editors and dozens of researchers.

The air-pollution group spent the entire academic year 1969-1970 at its task. The researchers were divided into three main teams, concentrating on (1) air pollution from local industrial activities; (2) the membership and activities of the Bay Area Air Pollution Control District (BAAPCD); and (3) the public reaction to air pollution, both on the man-on-the-street level and through organized citizens' groups.

The research team working on industrial air pollution studied twenty-nine Bay Area industrial sites in great detail with groups of researchers visiting twenty of them for a tour and interview. Additional information was obtained from BAAPCD files and other sources. (Although the private automobile is a major contributor to the Bay Area air-pollution problem, the workshop concentrated on industrial pollution instead. Air pollution created by cars is more a national than a local problem and has been much more extensively studied.) Their report contained detailed data on emissions, pollution-control achievements, and recommended improvements for each of these plants. A number of the plants studied were found to be seriously deficient—but a number of others were identified as exemplary.

The researchers studying the Bay Area Air Pollution Control District attended BAAPCD Board meetings and also many meetings of subsidiary councils and committees. They interviewed the directors and staff at length and studied the BAAPCD's public records. This information provided the basis for a thorough discussion of the history and organization of the BAAPCD and a cogent analysis of its accomplishments and shortcomings. In addition, the report of this group gave detailed information on each member of the board, each member of its influential Technical Advisory Panel, and the most important members of its staff. Overall, the report emphasized the BAAPCD's potential and urged citizens to help it become more aggressive by giving it their political support.

The final group of researchers conducted a public opinion survey. A total of 1,436 people were briefly interviewed at seventeen locations in the six-county Bay Area. Here are some typical responses:

"How serious is the air-pollution problem?"

Very serious	70.6%
Somewhat serious	25.3
Not serious	3.0
No problem	0.5
No opinion	0.6

Other questions established that most people would be willing to spend a significant amount of money (of the order of three to five dollars per month) in taxes or increased prices for cleaner air. However, only 10 percent knew who was responsible for regulating air quality in the Bay Area (the BAAPCD).

The researchers followed up their man-on-the-street survey with seventy-seven extensive telephone interviews with representatives of labor unions, men's and women's service clubs, church groups, and so forth. Finally, they reported on and evaluated the work of most of the local citizens' groups working for cleaner air and then gave detailed suggestions for individuals or groups interested in joining the fight.

The final product of this monumental effort was a comprehensive and remarkably readable 380-page handbook entitled *Air Pollution in the San Francisco Bay Region.*[5]

The SWOPSI air-pollution workshop concluded in spring 1970, and the report was relased the following September, along with a twenty-two-page summary. Television and other media coverage was good, and one San Francisco radio station, KCBS, quoted excerpts from the report and from a taped interview with Groth for several weeks afterward. Of the twenty-eight Bay Area daily newspapers seventeen covered the report, devoting an average of thirty column inches per paper to the story.[6] Unfortunately, none of the newspapers told their readers how they could obtain copies of the full report; this information was supplied only by San Francisco's noncommercial television station, KQED.[7] Nevertheless the demand for the report was high. More than 2,000 copies were distributed.

The report did not go unnoticed by the BAAPCD. A committee of the board was appointed to review it. When they reported back eight months later, however, all they had to say was that the report was basically sound and full of useful information but that, in their opinion, the section on the personalities of individual board members was in poor taste. Perhaps a more tangible response to the report occurred in August 1970, even before the report came out, when the board appointed Ned Groth to its Technical Advisory Panel. He replaced an industrial representative whose reappointment his group had strongly opposed. Thus, the first official reaction to the SWOPSI workshop was to coopt its leader.

In the years since the SWOPSI report, several older BAAPCD board members have been replaced with young activists, and the lobbying of citizens groups has become increasingly effective. Groth and his friends have given these groups assistance, including educating them on air pollution problems and organizing presentations by expert witnesses at BAAPCD hearings and in Sacramento.

Some Other SWOPSIs

PESCADERO DAM

One other of the first ten SWOPSI workshops had a considerable impact on local issues: a study of a proposed dam on Pescadero Creek, a pretty stream which winds through the mountains west of Stanford down to the Pacific Ocean. This workshop was led by J. D. Bjorken, a well-known theoretical physicist at the Stanford Linear Accelerator Center, and Joe Califf, an engineering graduate student specializing in water resources. The workshop found that the proposed dam was for geological reasons an exceedingly costly (about $50 million) way to supply water for proposed housing developments along the Pacific Coast south of San Francisco—itself a goal of arguable desirability. Furthermore, the dam would flood the central part of an important state park, and the reservoir thus created would be of limited recreational value because of large fluctuations in the water level. The workshop's report[8] and Bjorken's testimony were influential in convincing the county to abandon the project.

UNIVERSITY ISSUES

Several of the workshops concentrated on problems of special concern to Stanford University, its students, and its staff. One focused on helping graduating students find "jobs in areas of urgent social concern." Another studied problems in the delivery of health care in the United States, focusing particular attention on Stanford University's health care plans for students and employees. All of the six participants in this workshop were premedical students. One of their recommendations—which was adopted by the university—involved an improvement in the terms of Stanford employees' major medical insurance. Yet another workshop examined the impact of computers on privacy, studying both technical possibilities and desirable policies. As a result of a study of Stanford University's safeguards of student files by two participants in this workshop, the university instituted a number of reforms—some of them even before the report[9] appeared. In this case, as in others, the mere existence of a group studying the operations of the bureaucracy helped to provide the impetus for constructive self-examination.

NATIONAL ISSUES

Two of the first SWOPSI workshops attacked problems of national or international scope. One of these, led by the director of the Stanford Linear Accelerator Center, Wolfgang K. H. Panofsky, sought to find ways in which students could work for arms control. Professor Panofsky has had a great deal of experience as an arms control advisor and negotiator (See Chapter 5.)

More than 100 students sought to register for Panofsky's course. Although this was several times the number that could be accommodated, the students' obvious enthusiasm led Panofsky and several other faculty members to plan a large-scale course on arms control starting the following year. The SWOPSI workshop participants studied the problems of disarmament and diplomacy, and some helped to develop materials for the new course. Several of the students were selected to participate in an international summer school on arms control in Italy, and several others secured summer positions with the U.S. Arms Control and Disarmament Agency. One of Panofsky's assistants in the SWOPSI course, Elise Becket, then a second-year law student, went on to work in the summer of 1970 as an aide to Senator John Sherman Cooper (R.-Ky.), who was at the time one of the leaders in the Senate fight against the antiballistic missile system.

The authors of the present book were involved in yet another SWOPSI workshop—on federal policy making for technology. Other leaders of the workshop included Martin Perl, an experimental physicist, and Robert Jaffe, a graduate student of physics and also one of the organizers of SWOPSI.

A major focus of this workshop was a study of the federal science advisory system. Two former members of the President's Science Advisory Committee and several other high government advisors each spent an evening in discussion with the workshop participants, and several of the participants' research projects examined the role of technical advice in specific executive-branch decisions.

The group soon became concerned with the relatively weak role of Congress in determining national policy for technology. As one of the projects of the workshop, a questionnaire was sent to every member of Congress, with the cooperation of former Representative Jeffrey Cohelan of Berkeley and California Senator Alan Cranston (D.). The responses from eighty-two Congressmen indicated that most of them felt that Congress was at a serious disadvantage compared to the executive branch for lack of technical information and expertise. A small report, *Congress and Technology*,[10] was then written presenting the case for upgrading Congress's resources of technical expertise and giving particular suggestions as to how this might be done—among these a proposal for a program of Congressional fellowships for scientists. (Several professional societies organized such a program in 1973, as we describe below in Chapter 18.) The report was distributed to all members of Congress.

Another project of this workshop was a study of news media treatment of technical issues, in particular the oil leaks from wells in the Santa Barbara Channel, on the California coast. They found that almost all of the numerous articles on this subject in leading newspapers and news magazines were derived from official statements or handouts by government or industry, and only a very small fraction of the news coverage was based on investigative journalism.

During the summer of 1970, after the completion of the workshop, the present authors went on to write a 200-page report, *The Politics of Technology: Activities and Responsibilities of Scientists in the Direction of Technology.*[11] This report discussed the organization and effectiveness of the executive-branch science advisory structure. (Parts II and III of this book are an outgrowth of that project.) We were pleased but frankly astonished at the interest in the report when it came out. It inspired articles in publications ranging from *Chemical and Engineering News* to the *National Enquirer*,[12] and friends even sent us news clippings from England and Israel. Perhaps more importantly, it was rather widely discussed in the scientific community (and even by at least one panel of the President's Science Advisory Committee, where, according to an informant, the panel members were admonished by their chairman not to follow our suggested guidelines for advisors). It was the response to this report which convinced us that a book on the subject was required.[13]

Overview

The workshops discussed so far were among the first group of ten SWOPSIs offered in fall 1969. The program is still flourishing. During SWOPSI's first three years there were more than eighty workshops enrolling some 1,700 undergraduates and 200 graduate students at Stanford. More than half of the workshops have had an impact of one sort or other on the wider community, and over a dozen have prepared comprehensive and authoritative reports on various subjects, such as *Pesticide Exposure and Protection of California Farm Workers, The Politics of Pollution Control in Monterey Bay,* and *Balanced Transportation Planning for Suburban and Academic Communities.*[14] The transportation workshop in 1971 also produced a useful pocket-size handbook of public transportation in the Bay Area, *Ride On!,*[15] which is still selling well at local bookstores and newsstands.

The influence of the SWOPSI workshops has thus been considerable, both in the local political arena and in their effects on the participants' lives. Indeed, SWOPSI seems to be well on its way toward becoming a Stanford institution. Perhaps the most serious danger that the program faces is that it will become too "academic," overinstitutionalized—and less hard-hitting.

This is not to say that, to be effective, SWOPSI-type courses must be less academically oriented than traditional courses on traditional subjects. Indeed, the SWOPSIs complement the traditional curriculum. One of the greatest benefits of the SWOPSI approach has been in introducing students to the kind of field work that researching a social or political issue entails: isolating and structuring a research area, identifying and interviewing appropriate individuals, finding and securing relevant documents—frequently relatively obscure publications from government agencies or corporations. Workshop leaders have commented that undergraduates generally require a lot of initial guidance before they can successfully undertake such research. Enthusiasm often compensates for lack of experience, however, and students willingly pore over statistical data and learn to evaluate relevant chemical, engineering, and business techniques. The experience that they thus gain should be helpful in their future careers, and for some students it has influenced their choice of academic majors and career goals.

Faculty, too, are not immune from such influences, and SWOPSI workshop leaders have been able to develop new interests and apply knowledge and skills to fields that they would normally not enter. It must be admitted, however, that the successful SWOPSI workshops have made very heavy demands upon the time of their faculty and graduate student leaders. The leaders have not been compensated for their contributions to SWOPSI either in salary or by any reduction in their normal course load. Voluntary faculty support can sustain a new academic program through its experimental years, but it is unrealistic to expect it to continue indefinitely. Thus far the required large-scale funding has not been forthcoming from the government, from private foundations, or from

Stanford or other universities themselves. But SWOPSI has succeeded even without such funding—thanks to the dedication of its workshop leaders.

There is no reason why a program similar to SWOPSI cannot be instituted at any college or university of at least moderate size. The main requirements are a large measure of enthusiasm among some students and faculty and the willingness of a few people to organize it. It also helps to have some key administrators on your side. (In SWOPSI's case, the most helpful university official was Dean of the Graduate School Lincoln Moses.)

Political Constraints

A potential problem that worried the SWOPSI organizers even before the first workshops began was that persons outside the university would criticize the propriety of any university involvement in politics and challenge the objectivity of the workshop leaders and participants. For example, a skeptic might react to the logging workshop, described above, as an effort by Professor Cox to recruit undergraduates to fight his private battles. Actually, there has been little criticism of this type. This is probably due, at least in part, to the high quality of most SWOPSI reports as well as to the fact that most workshop leaders have been careful to restrict workshop activities to information gathering, analysis, and dissemination, with any political activity postponed until after the workshop has concluded.

The only attacks on SWOPSI have come from within the university, not from outside. In each case it was because some professorial oxen were gored. The most damaging of these attacks occurred after the publication of the two-volume SWOPSI report *Department of Defense-Sponsored Research at Stanford*.[16] Volume I simply reprints the statements on file at Stanford regarding the nature of the research being performed under each Defense Department contract, together with a computer printout from the Pentagon giving its version of the same information. Not surprisingly, in some instances the differences were pretty striking: the professor would claim to be doing some perfectly innocuous-sounding research project—for example, "High power broadly tunable laser action in the ultraviolet spectrum"—while the Defense Department report would emphasize the potential military applications of the same research: "Weaponry—lasers for increased damage effectiveness." Volume II of the report comments on these differences as well as on the more general implications of military sponsorship of university research.

The SWOPSI Policy Board had thought the report fair but dull. They were much surprised, therefore, to find it receiving considerable coverage in the news media.

The report was also greeted by cries of outrage from a number of the faculty members whose research it described. And, of course, university officials were concerned about the possible damage to the university's relationship with the Department of Defense. One day, when the controversy over the report was at its peak, a representative of the Stanford University research office called upon a leading official at the Pentagon in charge of Defense Department-sponsored research. The Stanford man wished to make it clear that he deplored the report, that he considered it irresponsible, and that the Stanford administration deeply regretted the whole affair. Much to his consternation, the Pentagon official disagreed, asserting that in his opinion the report was quite balanced—and that furthermore one of its authors was his daughter![17]

Unfortunately, the story did not end here. Some of the Stanford faculty, particularly certain members of the Stanford School of Engineering, brought strong pressure on the university administration to throttle SWOPSI. The Dean of Undergraduate Education, within whose bailiwick SWOPSI resided, responded by demanding better review procedures for SWOPSI publications. The university also refused to provide any support for the publications program, and it forbade SWOPSI to seek outside support.[18] The number of new SWOPSI publications subsequently declined sharply. The university's decision not to fund SWOPSI publications did have one virtue, however: by forcing the publications program to become self-supporting, it enabled SWOPSI to remain partially independent. By late 1973, several interesting new SWOPSI reports were in publication or preparation.[19]

NOTES

1. The undergraduate was Joyce Kobayaski, and the graduate students were Bob Jaffe and Joel Primack. Joyce served during academic year 1969-1970 as Stanford student body president and later became a medical student. Bob and Joel were graduate students in theoretical physics.

This book was stimulated by the authors' involvement in one of the first SWOPSI courses, discussed later in this chapter. And this chapter is based upon SWOPSI files, the personal records of the authors, and conversations and correspondence with Allan Cox, Edward Groth, Nicholas Corff, and Dan Lewis.

Copies of the reports mentioned in this chapter and more information on SWOPSI can be obtained from SWOPSI, 590A Old Union, Stanford University, Stanford, California 94305. See also Joanne Lublin, "Stanford's Recipe for Relevance," *Change, The Magazine of Higher Learning*, October 1971, pp. 13-15; Nicholas J. Corff, et. al, *SWOPSI Director's Report 1970-71;* and B. Michael Closson and James L. Gibbs, Jr., *A Report to the Senate of the Academic Council on the Special Joint Agencies of the Committee on Undergraduate Studies and the Dean of Undergraduate Studies,* Stanford University report no. SenD#1010, StCD#1436, November 1972.

2. Allan Cox and Davison Soper, *Logging in Urban Counties* (Stanford, Calif.: SWOPSI 1970).

3. The new law may not have been entirely responsible for the Santa Cruz Timber Company's demise. Informed local opinion is that the company would soon have

discontinued operations anyway: it was prepared to handle only virgin and old-growth stands, which had virtually all been cut down; and besides, it wanted to develop the land for other purposes.

4. Allan Cox, private communication, May 1971.

5. The Stanford Workshop on Air Pollution, Ned Groth, ed., *Air Pollution in the San Francisco Bay Area* (Stanford, Calif.: SWOPSI, 1970).

6. David M. Rubin and David P. Sachs, *Mass Media and the Environment: Water Resources, Land Use and Atomic Energy in California* (New York: Praeger Publishers, 1973), pp. 108-113, 272-275. Most of the newspaper articles stuck to generalities: only three of the seventeen papers discussed the report's evaluation of the performance of local industry in meeting air pollution standards, and only three reported on the evaluation of their local BAAPCD representative. (See also Ref. 7.)

7. Peter M. Sandman, "Mass Environmental Education: Can the Media Do the Job?," to be published in *Environmental Education*, William B. Stapp and James A. Swan, eds. (Beverly Hills, Calif.: Sage Publishing Company, 1974). Sandman points out that the news media's failure to tell how to get copies of the report is typical: "News stories are constructed so as to lead the audience to believe it knows all it needs to know."

8. James D. Bjorken and Joe Califf, *The Pescadero Dam and San Mateo County Coastside Development* (Stanford, Calif.: SWOPSI, 1970).

9. Greg Bomberger and Joyce Kobayashi, *Privacy and Student Records at Stanford University* (Stanford, Calif.: SWOPSI, 1970).

10. The SWOPSI Workshop on Technological Issues, Joel Primack and Frank von Hippel, eds., *Congress and Technology* (Stanford, Calif.: SWOPSI, 1970).

11. Frank von Hippel and Joel Primack, *The Politics of Technology: Activities and Responsibilities of Scientists in the Direction of Technology* (Stanford, Calif.: SWOPSI, 1970).

12. "Report Assails Science Advisory Agencies," *Chemical and Engineering News,* January 4, 1971, p. 22; "Govt. Suppressed Scientists' Warnings of Dangers to the Health and Safety of the Public," *National Enquirer,* February 21, 1971, pp. 1-2.

13. Martin Perl has also written on the problems of science advising: "The Science Advisory System: Some Observations," *Science,* 173 (1971): 1211-1215.

14. Christopher H. Lovelock, ed., *Balanced Transportation Planning for Suburban and Academic Communities* (Stanford, Calif.: SWOPSI, 1971; supplement published 1973).

15. *"Ride On!" The Stanford Guide to Public Transportation in the Bay Area* (Stanford, Calif.: SWOPSI, 1971).

16. Stanton A. Glantz, Carol A. Farlow, Richard A. Simpson, Norm Z. Albers, Dennis E. Pocekay, William Holley, Michael S. Becker, Stephen A. Ashley, and Michael R. Headrick, *Department of Defense Sponsored Research at Stanford—Vol. 1, Two Perceptions: The Investigator's and the Sponsor's* (Stanford, Calif.: SWOPSI, 1971); Norm A. Albers, Stephen S. Ashley, Michael S. Becker, Carol A. Farlow, Stanton A. Glantz, and Richard A. Simpson, *Department of Defense Sponsored Research at Stanford—Vol. 2, Its Impact on the University* (Stanford, Calif.: SWOPSI, 1971). See also Deborah Shapley, "Defense Research: The Names Are Changed to Protect the Innocent," *Science* 175 (1972):865

17. Private communication from Stan Glantz and SWOPSI Director Dan A. Lewis.

18. *SWOPSI Quarterly Report, Winter Quarter, 1972,* Stanford University report no. CUS 250-72, p. 10.

19. Brent Appel, Vance Peterson, and Jay Schoenau, eds., *The Other Stanford* (Stanford, Calif.: SWOPSI, 1973). Gordon Lewin, *Rapid Transit and the Public Interest: A case-study of the San Francisco Peninsula* (Stanford, Calif: SWOPSI, 1974).

CHAPTER 15

Challenging the Atomic Energy Commission on Nuclear Reactor Safety: The Union of Concerned Scientists

We had at the beginning of our work no inkling whatsoever that there was anyone within the depths of the nuclear community who shared anything like the positions we were developing.

As we continued to work and meet at Oak Ridge [National Laboratory] and [the National Reactor Testing Station], we were quite surprised to find the reactions of men there so close to ours. We found it personally astonishing once the hearing gained its momentum to see the number of people who were so clearly accepting a position quite divergent from the official position of the Atomic Energy Commission.

I think that this has been both personally to us and I think to the public at large one of the most revelatory aspects of this public proceeding.[1]

—Daniel F. Ford testifying at
the AEC hearings on reactor safety,
Bethesda, Maryland, August 22, 1972

Dan Ford

Dan Ford seems an unlikely person to trouble the powerful Atomic Energy Commission (AEC), much less the giant electric utility industry. Although still in his twenties, Ford, who has the deceptive appearance of an overgrown schoolboy, has become one of the key leaders in a movement to force a reconsideration of the country's rapidly increasing commitment to nuclear power for generating electricity.

Ford is not a scientist. He studied economics as an undergraduate at Harvard, obtaining his bachelor's degree in 1970. The environmental movement came into its own that year, and when Ford was offered the position of coordinator of environmental research for the Harvard Economic Research Project, where he had worked as an undergraduate, he jumped at the opportunity. The appointment offered a welcome pause before the academic routine of graduate school.

In his new job Ford was responsible for a pilot study on the costs and benefits of various methods of generating electrical power. The majority of electric power plants then being built in the United States were (and still are) nuclear, and there was considerable public controversy over the dangers of cancer and genetic defects from the small amounts of radioactivity which are released into the environment during the normal operation of these new plants—and also some concern about the possibility of a much larger release of radioactivity as a result of a serious accident at a nuclear power plant. It was natural, therefore, that Dan should look into these questions.

One day in the spring of 1971, while Ford was educating himself on nuclear reactors, he discovered that the AEC had published a notice in the *Federal Register* giving any interested public group thirty days in which to petition for a public hearing on the application of the Boston Edison Company for a license to operate its big new Pilgrim nuclear power station, located in Plymouth, Massachusetts.[2] The notice had received no press attention, and it seemed unlikely to Ford that there would be any response. He therefore decided to see to it that a hearing would be held so that the public could be informed and take action in its own interest.

Ford began by writing to Boston newspapers asking them to inform their readers about the AEC deadline. The only response was a brief article in the *Boston Globe*—but to Ford's exasperation, it failed even to mention the deadline. After several phone calls, Ford finally managed to convince the editor of the *Globe* to publish his letter just days before the deadline.[3]

The Union of Concerned Scientists

Meanwhile, however, Ford had received a response from another direction. A helpful reporter had put him in touch with Dr. James MacKenzie, a short, bushy-haired, energetic young physicist who had recently left MIT to work full time on environmental issues for the Audubon Society and who was also chairman of the MIT-based Union of Concerned Scientists (UCS).

The UCS had been organized in 1969 in response to the "March 4 movement" by student activists at MIT that challenged scientists to take public positions on the misuse of technology—particularly in Vietnam and the strategic weapons race.[4] Long after student activism had died down, the committed core of the UCS continued to work hard on issues they considered timely. Their initial focus was on preparing popular expositions of the technical arguments against new strategic weapons systems such as ABM and MIRV and against the Army's continued commitment to chemical and biological warfare. But a number of the UCS members became interested in the new political issues being raised by the environmental movement. When Ford contacted MacKenzie, the UCS was finishing a major study of the Boston air-pollution problem.

A meeting of the UCS was hastily called, and Dan Ford presented his case. He pointed out the disturbing fact that the AEC's Advisory Committee on Reactor Safeguards had expressed the opinion that the expected release into the air of radioactive fission products from the Pilgrim reactor would be excessive. There was also the question of the wisdom of placing the giant reactor so close to the Boston metropolitan area. Many experts—including AEC officials and the ACRS—had expressed the opinion that the barrier of distance is the most important protection for the general population in case of accidental release of some of the enormous store of radioactivity contained in a modern nuclear reactor. But the AEC had allowed the utilities to site reactors ever closer to metropolitan areas in order to reduce expenditures on power transmission lines.[5]

The UCS agreed that these were issues well worth exploring, and a small group decided to petition for a public hearing on the Pilgrim reactor. This was a significant commitment because, as a price for such an "intervention" in the licensing process, the AEC insists that any "intervenor" participate fully as a party in proceedings which sometimes drag on for many months. It testifies to the impression that Ford had made on the scientists of the UCS that they invited this young economist to join them and organize their participation in the Pilgrim reactor hearings. The petition was filed just hours before the AEC deadline.

The Battle over Nuclear Power

The UCS intervention was not an isolated action. The Pilgrim reactor was one of dozens which were being built around the country as the electrical utilities anticipated a rapidly growing national demand for electrical power. And the new nuclear plants, as the vanguard of a conspicuous new technology with a frightening potential for radioactive pollution, had become natural targets for environmental groups across the country.

The citizen groups that opposed the new power plants had been unable to find a respectable technical basis for their concern that the plants might blow up and spew lethal amounts of radioactivity over surrounding areas, so they were forced in licensing hearings to argue about more mundane problems. One of these was concern about increased cancer and genetic risks from the relatively low levels of radioactivity released from reactors during their normal operations. Two widely recognized experts, John Gofman and Arthur Tamplin of the AEC's Lawrence Livermore Laboratory in California, had articulated these concerns in a most forceful manner in articles, books, and testimony at reactor licensing hearings since 1969.[6] In 1971 the AEC retreated and proposed more stringent radioactive-release standards for nuclear reactors.[7]

A second major issue that had been raised by environmentalists is that of "thermal pollution" of lakes, rivers, and coastal waters. The amount of cooling water used by large modern power plants is enormous, with nuclear power plants requiring about 50 percent more than fossil fuel power plants of the same capacity because of their lower thermal efficiency and the fact that the fossil fuel power plants reject some waste heat through their smokestacks.[8] Starting in 1971, the utilities began installing cooling ponds and cooling towers costing millions of dollars.[9]

Emergency Cooling

Environmental groups had attacked nuclear power plants both on the basis of their everyday releases of small amounts of radioactivity and because of their thermal pollution—and had been substantially appeased. But just as the Union of Concerned Scientists was entering the fray, the unspoken issue—the danger of catastrophic releases of large amounts of radioactivity—finally surfaced.

Early in their preparation for the Pilgrim reactor hearings, the UCS contacted citizens' groups engaged in similar interventions in connection with other reactors. From the Businessmen for the Public Interest, a Chicago group which was supporting interventions into the licensing of a number of nuclear reactors around Lake Michigan, they learned of the failure, in semi-scale-model tests, of a crucial reactor safety apparatus known as the "emergency core-cooling system."[10] The tests had been performed at the AEC's National Reactor Testing Station in Idaho in November and December 1970.[11] Dan Ford, Jim MacKenzie, and two other UCS scientists—Ian Forbes and Henry Kendall—decided to educate themselves on the purpose of the emergency core-cooling system (ECCS) and the consequences if it failed to work as designed.

A typical reactor of the sort now being licensed for operation generates about a billion watts of electricity—enough electrical power to supply the needs of nearly a million Americans. This power originates in the heat generated by the splitting ("fissioning") of uranium nuclei in the reactor "core." The core, typically about twelve feet long and fourteen feet in diameter, contains about a hundred tons of uranium formed into ceramic pellets of uranium oxide held inside tens of thousands of long, thin "fuel rods." At full power the energetic fission fragments heat up the centers of the uranium oxide pellets to about 4,000° F ("degrees Fahrenheit"). The heat flows out through the zirconium alloy "cladding" of the fuel rods to heat up the high-pressure water circulated between the fuel rods. In carrying away the heat to power turbines that generate the electricity, the circulating water keeps the fuel rod cladding at the relatively low temperature of about 600° F. If the water were to be lost—through a broken pipe, for example—the cladding would heat up and rupture unless the emergency core-cooling system (ECCS) could reflood the core with water within a minute or so. This would occur even though the "chain reaction" stops as a result of the loss of the neutron-slowing action of the water which has been operating for some time the fission products build up to the point where their radioactivity alone generates enough heat to melt the core.

If the ECCS in such a reactor for any reason failed to do its job adequately when called upon, the ensuing events would be dramatic.[12] Within minutes the core, with its hundred tons of uranium, would begin to melt from the heat of its intense radioactivity and slump to the bottom of the reactor vessel. By this time the situation is already beyond control. Any attempt to cool the molten mass would only exacerbate the problem: the water would react with the hot metal chemically, liberating still more heat and explosive hydrogen gas. Within an hour the molten core would melt through the six-inch thick steel reactor vessel, releasing an immense amount of radioactivity, equivalent to the fallout from a large number of Hiroshima-sized nuclear bombs, into the reactor containment chamber—the domed concrete shell within which the reactor vessel and its primary cooling system are housed. Despite its name, the containment chamber would also be unable to keep the seething core from reaching the human environment. About a tenth of the core's total radioactivity is in the form of radioactive gases. Chemical explosions might occur, causing the containment shell to crack open and releasing these gases into the atmosphere. Even if the dome remained intact, the core would melt through the concrete floor of the containment chamber within about a day and would continue to melt its way down through the earth and rock below—probably for hundreds of feet. Because of the path that the core takes in this scenario, it is half-jokingly called the "China syndrome."

A location hundreds of feet underground might at first sight seem to be an ideal final resting place for the intensely radioactive core. Unfortunately, there is no guarantee that much of the radioactivity would still not escape to the surface. The hot radioactive gases could seep up into the air (if they had not already done so), and the remainder of the core would be available to contaminate the ground and surface water. Because of the enormous thirst of nuclear reactors for cooling water,[13] they are generally built on riverbanks, lakeshores, or seashores. The contamination of these waters could be on a very large scale. There is sufficient long-lived radioactive strontium-90 in a large reactor core to contaminate thousands of cubic miles of water—many times the volume of Lake Michigan—to a level greater than that which the AEC considers safe.

The UCS scientists extrapolated a 1957 AEC analysis[14] of the possible consequences of a hypothetical reactor accident in which a large fraction of the core's radioactive gases were released into the atmosphere to apply to the much larger reactors then coming into operation. Their conclusions beggar the imagination:

> If . . . the radioactive materials are released under a temperature inversion, by no means an uncommon nocturnal condition, with a 6.5 mph wind, . . . lethal effects can extend 75 miles downwind in a strip of maximum width up to 2 miles. Injuries would be likely at up to one or two hundred miles, the presence of moderate rain yielding the lower figure. . . .
>
> . . .The cloud would be increasingly difficult to see after it had moved away from the accident site, and would be invisible long before it had lost its lethality.[15]

Nearby cities would have to be evacuated as rapidly as possible. Long-term restrictions on normal use of the contaminated area would be inevitable. According to the UCS scientists, such restrictions would extend a minimum of fifteen miles from the reactor site and could reach distances of hundreds of miles.

Summarizing the implications, the UCS authors concluded:

> It is abundantly clear from our study that a major nuclear reactor accident has the potential to generate a catastrophe of very great proportions, surely greater than any peace-time disaster this nation has ever known. The full scale and consequences of such a catastrophe cannot fully be recorded, yet it is against such an ill-understood but awesome event that the scale of, and confidence in, the reactor safeguards must be weighed.[16]

Despite its tremendous importance, the emergency core-cooling system was designed almost entirely on the basis of greatly simplified computer calculations. The purpose of the semi-scale tests at the National Reactor Testing Station in Idaho was to verify that these computer programs in fact correctly simulated the behavior of the ECCS. But when the tests were actually conducted, the results were not as expected. The model did not behave as the computer programs had predicted. Instead of cooling off the model reactor core, the emergency cooling water was swept away by the escaping steam out the same pipe break through which the original cooling water escaped. Since the model was not realistic the failure of the tests reflected most directly on the computer programs, but the predicted effectiveness of the emergency cooling systems of actual reactors is based on such programs. The AEC has scheduled a much more elaborate series of "loss-of-fluid tests" in Idaho starting in 1974. In the meantime, the AEC and the reactor companies have been working at improving the computer programs.

Perhaps the most puzzling thing about the Idaho tests was their timing. The nation's utilities were investing tens of billions of dollars in nuclear reactors, but the AEC had given its safety program such a low priority that it had hardly begun testing the effectiveness of the emergency cooling systems of these reactors by the time they were being frozen in steel and concrete. The AEC's "watchdog" Advisory Committee on Reactor Safeguards (ACRS) had urged for years that these safety systems receive realistic tests, but the Idaho "teakettle-sized" tests represented only the first, extremely unsophisticated steps in the testing program.[17] Although the significance of the Idaho tests remains debatable, the AEC's irresponsible neglect of its reactor safety program could hardly be disputed.

The UCS issued its report[18] on the possible implications of the failure of the Idaho tests at a press conference in July 1971. This was the first such discussion intelligible to the layman, and it caused a sensation. That same evening both the NBC (Huntley-Brinkley) and CBS (Cronkite) network news programs reported the story on nationwide television. A new national controversy had been born.

To License or Not to License?

The AEC faced a real dilemma in the area of reactor licensing: How could it issue operating licenses for nuclear power plants when the Idaho tests had raised a serious question as to their safety? At the beginning of May 1971 AEC Chairman Glen Seaborg wrote the chairman of the Congressional Joint Committee on Atomic Energy, Senator John O. Pastore (D.-R.I.), telling him that the AEC expected delays in reactor licensing while a "senior task force" reviewed emergency cooling system effectiveness.[19] Then on May 13, 1971, AEC officials appeared before the Joint Committee requesting additional funds for reactor safety research. AEC Assistant General Manager for Reactors George Kavanaugh acknowledged that the test results were causing concern: "If [the situation] were better, we might not have been allowed to come up here asking for money."[20]

Meanwhile, every month's delay in starting up their new reactors would cost the electrical utilities millions. They would not sit patiently by awaiting the results of a long, drawn-out review—especially since it was generally agreed that the probability was remote that any particular reactor would suffer an accident serious enough to strain the capabilities of its emergency cooling system. It was out of the question to wait on the results of a comprehensive testing program. That would take years. A few weeks after its formation, therefore, in mid-June 1971, an AEC task force came up with proposed new "Interim Acceptance Criteria" for reactor emergency cooling systems. Although the recommendations lacked supporting documentation, the collective leadership of the Atomic Energy Commission, the five AEC Commissioners, quickly accepted them and promulgated them formally on June 29, 1971.[21]

In fact, the Interim Criteria were remarkably convenient for the nuclear industry. All reactors already operating or up for licensing could satisfy them. If the criteria had been much more stringent, the reactors might well have been required to operate at much lower and correspondingly less economical power levels or to undergo major modifications. Either course of action would have been a disastrous blow to the prestige of the AEC—possibly even to the future prospects of nuclear power.

As far as the AEC was concerned, the matter was settled and reactor licensing could proceed. The UCS group was not so sure—but what more could they do? No great expertise had been required on their part to draw the public's attention to the failure of the Idaho tests or to cite the 1957 AEC report on the possibilities for catastrophe should a major release of radioactivity occur in an actual reactor accident. But to oppose the Interim Criteria would be to challenge directly the technical judgment of the Atomic Energy Commission. The AEC had accumulated an enormous reservoir of expertise during the quarter-century of research and development which had gone into the design of the latest generation of commercial nuclear reactors, and the UCS scientists were quite unfamiliar with reactor engineering. In fact, Henry Kendall, who was to become the chief technical expert of the UCS in this area, later admitted that, at the time of the original UCS report on the emergency cooling problem, he was uncertain about even the most basic design differences between the two major types of commercial water-cooled reactors.

On the other hand, the AEC's case for the adequacy of its ECCS Interim Criteria could be no stronger than its weakest link. The challengers would not have to match the full range of expertise available to the AEC in order to challenge the AEC's conclusions. Furthermore, while the engineering details of nuclear reactors might be unfamiliar to the UCS scientists, the physics was not. They were confident in their abilities to understand quickly the calculations on which the Interim Criteria were based. So, minus their one nuclear engineer (Ian Forbes, who had to return to full-time teaching at the Lowell Technological Institute in Massachusetts), Kendall, Ford, and MacKenzie started to study the AEC analysis.

They soon found that, in the absence of actual experimental information on how an emergency core-cooling system might work, the AEC task force had again relied on highly simplified mathematical descriptions of the reactor and core-cooling system, with the ECCS performance being predicted using computer simulations. But the "garbage in-garbage out" axiom of computer experts seemed to the UCS scientists to be highly relevant here. Not only were the computer models necessarily oversimplified in the face of the complexity of the phenomena occurring in a nuclear reactor which had just lost its cooling water, but also, the UCS scientists found, crucial assumptions had been made that were demonstrably false.

One of these assumptions—that the geometry of the reactor core would remain unchanged during a loss-of-coolant accident—was directly contradicted by the results of experimental tests on fuel rods at the AEC's Oak Ridge National

Laboratory in Tennessee. In these tests the fuel rods began to swell, buckle, and rupture at temperatures hundreds of degrees lower than the peak temperatures that the AEC's Interim Criteria specified as allowable in the interval during which the core would be uncovered by water.[22] The AEC canceled its funding for these crucial Oak Ridge experiments in June 1971 (soon after the troublesome results began to appear), eliciting the following protest from Oak Ridge's director of nuclear safety research, William B. Cottrell: "We are astounded at your decision to discontinue this experimental work. . . . No one really knows what will happen in a reactor core in the event of a loss-of-coolant accident."[23]

The UCS group issued a report detailing its criticisms of the AEC's Interim Criteria in October 1971.[24]

The Licensing Hearings

The UCS attempt to challenge the adequacy of the emergency cooling system of the Pilgrim nuclear reactor at Plymouth, Massachusetts, in 1971-1972, was opposed by Boston Edison, owner of the reactor, on the grounds that the installation conformed to the AEC's Interim Criteria.

The emergency cooling issue had meanwhile been injected into licensing hearings on several reactors in other states. In November 1971 Dan Ford participated as a technical interrogator on this issue in the hearings on the Indian Point 2 reactor, located on the Hudson River above New York City. After much deliberation, the Indian Point 2 hearing board was at least partially persuaded by the UCS case, and in December 1971 it informed the AEC that it had serious questions about both the technical and the legal validity of the Interim Criteria.

In order to avoid further challenges on emergency core cooling in hearings on individual reactors, the AEC decided to hold comprehensive national hearings on this subject. The AEC initially proposed that these hearings be merely "advisory." But after negotiations with lawyers representing the Consolidated National Intervenors—a newly formed coalition of environmental groups which had been involved in individual reactor licensing hearings—the AEC agreed to rule on the emergency cooling issue on the basis of the record established in these "rule-making" hearings. Information possessed by the AEC, the reactor manufacturers, the electric power companies, and the Intervenors would be placed in the hearing record and subjected to cross-examination. Nevertheless, the AEC steadfastly refused to allow the Intervenors to subpoena documents or individuals, a right which had always been accorded to all participants in local nuclear reactor interventions.

The AEC's Experts

In order to get additional information, Dan Ford, Henry Kendall, and Jim MacKenzie had decided in November 1971 to pay a visit to the AEC's reactor safety experts at Oak Ridge National Laboratory. The UCS group was surprised to find that the Oak Ridge scientists were generally in agreement with their own misgivings about the AEC's new regulations on reactor safety systems. They returned to Cambridge laden with useful AEC documents. After the visit, the laboratory's Associate Director, Donald Trauger, reported to Milton Shaw, director of AEC's Division of Reactor Development and Technology:

We felt that the technical publication of this group, as well as their professional integrity, justified our meeting with them. However, inasmuch as the Union of Concerned Scientists has intervened in the hearing on the . . . Pilgrim reactor, we also felt that you should be aware of the nature of our discussions. . . .
. . . H. W. Kendall . . . showed us how he has used our data . . . to demonstrate that approximately 85 percent of the fuel rods are "candidates" for producing . . . coolant channel blockage in the range 70 to 100%. Kendall had reached this conclusion independently, and wanted to know if he was using our data properly—which he was, within the limits of its accuracy. . . .
The three members of the Union of Concerned Scientists who visited here appeared to be well educated and dedicated people. . . . They have become intimately familiar with the relevant published literature. . . . They have become aware of various deficiencies in the case for ECCS performance.[25]

The UCS group had already begun to acquire an extensive library of AEC documents on reactor safety. Their first major acquisitions were documents picked up, at Dan Ford's request, by an MIT physicist visiting Oak Ridge in June 1971. These were supplemented by documents obtained by Ford on a trip to AEC headquarters in Germantown, Maryland, as a special consultant on environmental economics in July 1971. (The AEC documents in question were not widely distributed, but they were not secret. The entire U.S. civilian reactor program has been unclassified for many years.) With their trip to Oak Ridge, however, the UCS had for the first time acquired access to an even more valuable source of information: they had won the confidence of some of the people who *wrote* the AEC reports. Ford and Kendall followed up their visit to Oak Ridge with trips to the Battelle Memorial Institute in Columbus, Ohio, an AEC reactor safety contractor, and to the AEC's National Reactor Testing Station in Idaho, where the scale-model cooling tests had been done. Although the officials at these institutions were less cooperative than those at Oak Ridge, Ford and Kendall found the scientists there not reluctant to discuss their own work.

The Hearings Begin

They have opened up a Pandora's Box of scientific doubts and bureaucratic heavy-handedness.[26]
                                                              —Nucleonics Week

The AEC's hearings on reactor emergency safety systems began in January 1972 with several days of legal wrangling between Myron Cherry, one of the lawyers representing the Consolidated National Intervenors, and the hearing board. At issue were objections to Dan Ford's participation in the hearings as a "technical interrogator" for the Intervenors and the Intervenors' demands that a number of AEC internal documents be put into the record and that representatives of the AEC's Advisory Committee on Reactor Safeguards be available at the hearings for questioning.

Ford's participation was objected to both by the board and by lawyers representing the reactor manufacturers and the electric utility companies on the grounds that he was not technically qualified. Ford admitted that he had never studied physics in college but maintained that he nevertheless could "ask the right questions" in the hearing as a result of his work with the UCS scientists. Although one of the hearing board members criticized him as an "instant expert," the board eventually decided to let him participate on a provisional basis "since Ford is the best the National Intervenors say they can produce."[27] (Teaching and research responsibilities prevented the scientific members of the UCS reactor safety team, Professor Kendall in particular, from participating regularly in the hearings.)

The AEC ruled that its Advisory Committee on Reactor Safeguards would not be represented at the hearing, and it refused to divulge that committee's formal review of the Interim Criteria on emergency cooling. Threatened with a lawsuit under the Freedom of Information Act, however, the commission decided to overrule the hearing board and released most of the other documents demanded by Cherry. Most of these were AEC staff memos concerning the emergency core-cooling system. This decision was an important windfall for the Intervenors, for these documents revealed the existence among the AEC staff of a great deal of uncertainty about the effectiveness of the reactor safety systems and the adequacy of the Interim Criteria.

The hearings had opened with considerable fanfare in a plush auditorium at AEC headquarters. But as the Intervenors began to hammer away at the AEC's case, the proceedings were moved to a rented office building in Bethesda, Maryland. Armed with the just-released internal memoranda, Cherry and Ford began to undermine the confident façade presented by official AEC witnesses.[28]

In a memorandum of June 1, 1971, less than a month before the Interim Criteria had been issued, the Chief of the Systems Performance Branch of the AEC's Division of Reactor Standards, Dr. Morris Rosen, and his deputy, Robert J. Colmar, had sent their final detailed criticisms of the developing Interim Criteria to the AEC task force charged with preparing them. Rosen and Colmar disagreed not just with small details of the proposed criteria but with the entire logic behind them:

The [AEC Division of Regulation] task force has undertaken to resolve the current regulatory difficulties . . . by attempting to formulate a "prescription" to be applied to each reactor vendor's codes and to be used as a basis for licensing reactors on a plant-by-plant basis.

This approach is predicated on the notion that the codes in their present state of development are definitive. . . .

We take exception to this current approach. We have consistently pointed out that this approach is too limited for the task at hand. . . . We believe that the consummate message in the accumulated code outputs is that the system performance cannot be defined with sufficient assurance to provide a clear basis for licensing.[29]

(In the AEC argot, "vendor," though it may conjure up images of Coke machines, actually means Westinghouse, General Electric, or one of the other reactor manufacturers; and a "code" is nothing more exciting than a computer program used to calculate phenomena such as the temperature of the reactor fuel rods during an accident.)

When Rosen and Colmar eventually were allowed to testify at the hearings, they expressed their misgivings about the AEC's reactor safety-system standards in even stronger terms. Rosen presented an eighty-page critique of the Interim Criteria. He said that he was disturbed and discouraged

to continue to see the advice of what I believe can be considered a significant portion of, more likely, a majority of the knowledgeable people available to the Regulatory staff, still being basically disregarded. . . .[30]

Margins of safety once thought to exist do not, and yet reactor power levels continue to increase resulting in an even more tenuous situation.[31]

Colmar explained in his testimony how he had become aware of the deficiencies in emergency cooling systems as early as February 1970—nearly a year before the Idaho scale-model tests—in the process of correcting Westinghouse's misinterpretation of its own computer programs. He stated flatly that in his opinion some form of reduction in reactor operating power was desirable until more experimental information on the effectiveness of reactor safety systems became available and characterized the Interim Criteria as "a triumph of hope over reason."[32]

Early in January 1972, Rosen was removed from his job and given an advisory position, and Colmar requested a transfer. Rosen was philosophical about the switch, saying that he had to "consider it as a promotion. . . . Of course, I am off ECCS—except in an overlook position."[33] He was also quoted as saying that "it's the sort of thing that, if it happened very often in an organization, you'd have to wonder."[34] Later he left the AEC.

G. Norman Lauben, one of the members of the AEC task force, had served in Dr. Rosen's department. When the Intervenor's lawyer, Myron Cherry, inquired during the hearings whether any of the task force members present could not personally support the official testimony, Lauben reluctantly raised his hand. He explained that if a certain variable in the computer programs used to evaluate the cooling system's effectiveness were decreased by as little as 20 percent—an amount that others testified was within the uncertainty of measurement—then reactor emergency cooling systems deemed acceptable under the Interim Criteria might actually be unable to prevent catastrophic core meltdown. In Lauben's opinion, insufficient experimental information was available to justify this lack of conservatism on the part of the AEC. When the chairman of the AEC task force, Dr. Stephen Hanauer, disagreed and claimed that the Interim Criteria were adequately conservative, Cherry asked:

CHERRY: Dr. Hanauer, in the area in which [Lauben] stated that he thought that the codes ought to be more conservative, can you state, sir, whether you believe that in that area you or Mr. Lauben possesses a greater understanding of the problem, in your judgment.
HANAUER: I think Mr. Lauben does.
CHERRY: Thank you, Dr. Hanauer.[35]

Hanauer later also admitted under cross-examination that three of the members of the AEC Advisory Committee on Reactor Safeguards who were most knowledgeable in the area of emergency core cooling had expressed concerns about the adequacy of the Interim Criteria.[36]

The reluctance of AEC witnesses to express open criticism of their superiors is understandable. But throughout the hearing the Consolidated National Intervenors continued to receive many letters, reports, and memos in addition to those officially released by the AEC. "The AEC leaks like a sieve," remarked Cherry cheerfully in explaining that many of the documents arrived in the mail in unmarked envelopes.[37] One of the most revealing of these documents, labeled "Hints At Being a Witness," was obtained in another way, however: it was accidentally given to Dan Ford by one of the AEC legal staff. Hint number 10: "Never disagree with established policy."[38]

Having finished the cross-examination of the AEC task force, the Intervenors next questioned the scientists from Oak Ridge National Laboratory whom the UCS team had met on their visit several months earlier. One of these witnesses was the scientist whose important work on fuel rod failure had been terminated abruptly by the AEC in June 1971, Oak Ridge metallurgist P. L. Rittenhouse. His written testimony was bland enough to satisfy the AEC bureaucracy, but when Rittenhouse actually appeared at the hearings to defend his testimony, he was sharply critical of the Interim Criteria. Under questioning by Cherry, Rittenhouse jolted the proceedings by asserting that a great many of his colleagues in AEC laboratories and the AEC headquarters staff shared his concerns. When asked to back up this assertion, he pulled out a list of twenty-eight names which he proceeded to read into the record. He described these individuals as persons "whom I have worked or at least talked with personally more than once. . . . These people have too many reservations

... shared too generally for me to pass off."[39] One of those whose name was mentioned by Rittenhouse was William B. Cottrell, Director of the Nuclear Safety Program at Oak Ridge. The Intervenors introduced into the hearing record a long letter, replete with supporting documents, from Cottrell to AEC headquarters:

To summarize what follows herein, we are not certain that the Interim Criteria for ECCS adopted by the AEC will, as stated in the Federal Register, "provide reasonable assurance that such systems will be effective in the unlikely event of a loss-of-coolant accident."[40]

Shortly after Cottrell had sent this letter, in December 1971, one of his superiors at Oak Ridge had called AEC headquarters asking that the letter be returned, claiming that it was only a "draft." Subsequent testimony by Cottrell established that the letter did in fact represent the views of a number of Oak Ridge reactor safety experts and that it was not a draft.[41]

In the first months of the reactor safety hearings, the Intervenors thus disclosed a deep rift between the AEC's reactor experts—particularly those in the AEC's laboratories who studied reactor safety problems—and the AEC bureaucracy, who channeled funds for the research and had to act on the results. The extent of the resulting tension will perhaps be indicated by the following remarkable letter from Alvin Weinberg, director of Oak Ridge National Laboratory, to James Schlesinger, Chairman of the Atomic Energy Commission:

Dear Jim:

When you called me in Florida you asked me to make clear to our [Oak Ridge] people that, when they testify at the ECCS hearings, they are to present their views fully and without reservation. I have conveyed this message to Messrs. Rittenhouse, Trauger, Cottrell, [and others]. That some of the testimony may prove to be in conflict with the interim criteria will not prevent them from presenting their data and conclusions as honestly and fairly as they can.

With respect to the criteria themselves, I have only one point to make. As an old-timer who grew up in this business before the computing machine dominated it so completely, I have a basic distrust of very elaborate calculations of complex situations, especially where the calculations have not been checked by full-scale experiments. . . . This is expensive, but there is precedent for such experimentation—for example, in the full-scale tests . . . on nuclear weapons.

I have one other point. I believe [Oak Ridge] and the other National Laboratories should have been as intimately involved in the preparation of the interim criteria as we have since been in the preparation of AEC testimony for the hearings. That we were not so involved reflects a deficiency in the relation between Laboratory and Commission that troubles me. . . . [The AEC's National Laboratories] must be called upon fully by the Commission even when this may uncover differences of opinion between the Laboratories and the staff of the Commission. . . . I can guarantee that our opinion, if solicited, will be both honest and responsible.[42]

National Intervenors vs. Milton Shaw

Barely concealed behind the diplomacy of Weinberg's letter was a history of steadily worsening relations between the National Laboratories and the AEC Division of Reactor Development and Technology, headed by Milton Shaw. During his eight years as the AEC's reactor czar, Shaw had won a reputation as a hard-boiled engineer and an autocratic administrator. His empire included not only all AEC design and development of conventional and "breeder" reactors (reactors which would convert enough non-fissionable materials to fissionable fuel to more than replenish the fissionable fuel which they "burned"), but also all reactor safety research. Shaw's office was thus in a position to curtail crucial research on the safety systems of commercial nuclear reactors; to censor unwelcome reports—even to impede the communication between the safety experts and the AEC Regulatory staff which is responsible for certifying the safety of the designs of commercial reactors; and to intimidate or transfer dissenting AEC employees. There is evidence that Shaw's office actually did each of these things.

In a series of articles, *Science* magazine reporter Robert Gillette documented the continued neglect by the AEC of crucial safety research. Indeed, Gillette indicated that Shaw's office had even gone so far as to spend money authorized for safety studies of conventional reactors on the development of the breeder reactor instead. Gillette reported that from 1965 through 1968, $12 million, or 8.5 percent of the funds appropriated by Congress for reactor safety research, were diverted to other purposes or simply not spent. And of the money actually spent for reactor safety, development of safety systems for future breeder reactors cut sharply into expenditures for safety research for ordinary reactors.[43]

The fate of a detailed report on emergency core cooling research needs prepared by the staff of the National Reactor Testing Station (NRTS) in Idaho (part of Shaw's command) illustrates the blockage of the AEC's internal communication channels. On April 4, 1972, the Intervenors had the opportunity to cross-examine J. Curtis Haire, manager of the nuclear safety program of Aerojet Nuclear, the AEC's primary contractor for light water reactor safety research at the NRTS. Haire admitted that his laboratory's reports on nuclear safety were sent to Shaw's office for review prior to publication and that, in its reports on the failure of the semiscale-model emergency cooling tests, Aerojet had been forced virtually to eliminate discussions of the relevance of these tests to the effectiveness of emergency cooling systems. The next day Shaw himself happened to be on the witness stand, and Cherry asked him if it was not a fact that the Idaho reports were being censored and edited. Shaw replied:

Censoring? If you want to use that terminology in the sense I think you are using it, yes. . . . I think it is a basic requirement that reports that are issued by people who are working for us have in them factual information, they are not speculative in the sense of not referring to things they should not.[44]

Haire was then questioned again the following day.

CHERRY: Now is it a fact, Mr. Haire, that the censoring which is going [on] . . . is not a disagreement with . . . technical judgement, but, rather, results in an inhibition of a free and open discussion of [the NRTS] views on safety?

HAIRE: Yes, it is rather an inhibition of free and open discussion rather than a matter of taking issue with technical matters. . . . I believe that RDT [Shaw's Division of Reactor Development and Technology] is trying to avoid the problem or burden, if you will, of having to spend a lot of time answering public inquiries that are addressed to them.

CHERRY: On nuclear safety?

HAIRE: On general questions of nuclear safety, yes.

CHERRY: Now, sir, this belief, is it based on any conversations with persons at RDT?

HAIRE: Yes.

CHERRY: Who?

HAIRE: Mr. Pressesky [Andrew Pressesky, Shaw's deputy for reactor safety].

CHERRY: He told you that?

HAIRE: In substance, yes.[45]

Curtis Haire was subsequently removed from his job and given a position in charge of "program development." It was of course denied that this action was taken in reprisal for his testimony.[46] But Haire's boss, the president of Aerojet Nuclear Corporation, had warned that if any employee's comments

sour his relationship with the customer [the AEC], we cannot guarantee that after some time has elapsed he will still be in his same position. We would, however, make every effort to find him a suitable opening.[47]

A similar rule was put into effect at Oak Ridge National Laboratory at the insistence of Shaw, and so Oak Ridge director Weinberg was able to protect the jobs of employees in ill favor with the reactor czar only by transferring them to a part of the laboratory not within Shaw's jurisdiction.[48]

Shaw's handling of the nuclear reactor safety program became one of the key points of contention on the "hidden agenda" of the reactor safety hearings. There was therefore great interest when he finally took the stand himself. Dan Ford was the technical interrogator.

In the course of his testimony and Cherry's preliminary cross-examination, Shaw consistently maintained that the Interim Criteria were adequate. He professed to be entirely unshaken in this conviction by the adverse testimony presented at the hearings, and he even asserted that no important experimental data were lacking in support of the criteria. Shaw also maintained that he had prepared his written testimony entirely by himself and took full personal responsibility for the judgments expressed therein.[49] Since these judgments were at such variance with those offered by the experts from Oak Ridge and Idaho, Ford pressed Shaw to back them up:

FORD: Mr. Shaw, I would like to ask you some questions about page 22 of your testimony and your opinion that one of the major areas of conservatism is related to the area of blowdown heat transfer. Now with respect to blowdown heat transfer, the interim criteria [use the] Groeneweld correlation. Can you tell me, Mr. Shaw, is the Groeneweld correlation a steady-state or a transient heat transfer correlation?

SHAW: I would prefer to cover this as I indicated before. [Shaw had earlier asked for an opportunity to consult sources before replying to questions.] . . .

FORD: Did you ever know whether it was a steady-state or a transient heat transfer correlation?

SHAW: I cannot recall whether I ever addressed this question in those terms.

FORD: Have you ever read the [AEC report] referenced in the interim policy statement as the source of the Groeneweld correlation?

SHAW: I cannot recall whether I ever read that document. . . .[50]

FORD: What are the documents you consulted?

SHAW: Mr. Ford, I have been in this business twenty-some-odd years. All right? The information relating to this goes back through these years. My job depends upon this information over these twenty-odd years. I cannot recall every bit of information that I used in this regard nor do I see any good reason to try to do it.

FORD: What documents did you consult?

SHAW: I do not recall.

FORD: Do you not recall any?

SHAW: I do not recall the documents. I am sure I depended a great deal upon my background.[51]

Ford emphasized to the hearing board that his queries were not "curve-ball or esoteric questions . . . thought up just to test the witness." They were questions on the basic literature, on subjects and references that played an important part in the Interim Criteria. The questioning continued:

FORD: Well, what is the basic experimental source of information on reflooding heat transfer, Mr. Shaw?

SHAW: Again, I believe that is detail, if you don't mind. . . .

FORD: Have you ever heard of the FLECHT program?

SHAW: Oh, absolutely. In fact, I think I initiated it, didn't I?

FORD: But you did not seem to recall that the FLECHT program was the basic source of experimental data on heat transfer in the reflooding period. How in the world do you explain that?

ENGELHARDT [AEC chief counsel]: I object to that, Mr. Chairman. It is argumentative.

CHAIRMAN GOODRICH: I will sustain the objection, much as I would like to hear the answer.

(*Laughter*)[52]

The nuclear industry press was uniform in its opinion of the outcome of the day's hearing. *Nucleonics Week,* a McGraw-Hill trade paper, stated its impressions as follows:

Milton Shaw, director of the AEC Div. of Reactor Development and Technology and thus head of the government's civilian nuclear power program, was verbally floored by the National Intervenors last week at the rulemaking hearing on emergency core cooling. In a theatrical day of questioning, Shaw simply was unable to answer direct questions about his own written testimony, although he maintained over and over again that he was indeed the author of that testimony.[53]

Shaw's disastrous performance at the ECCS hearings may have been more a reflection of his basic lack of interest in the safety problems of conventional reactors than of any lack of ability. It was widely believed in the AEC's National Laboratories that Shaw had one overriding ambition: to be known as "the father of the breeder reactor." This ambition is in accord with the tradition of the AEC. As one critic of the agency has said:

In any technical adventure, there are exciting parts and there are dull parts. An analysis of every AEC blunder to date indicates clearly that the AEC has accomplished the exciting aspects of every job with competence, expertise and dispatch. But as with individuals, organizational competence isn't defined as doing exclusively just what pleases and satisfies. There's also the dull but inescapable part of any job which must get done, too, like cleaning up the mess after a job is over.[54]

Ford and Kendall Cross-Examined

Although at the opening of the hearings the nuclear industry dismissed the dissent within the AEC over the Interim Criteria as "healthy,"[55] Shaw's humiliation and the accumulating weight of expert testimony against the Interim Criteria soon forced a reassessment. "ECCS Situation Growing Steadily More Ominous for AEC, Industry," headlined *Nucleonics Week* on April 20, 1972. The accompanying article reported that, in a meeting of the AEC Commissioners with top staff officials, including Shaw, there had been

"hard questions" on how AEC had gotten into its present position. . . . AEC chairman James Schlesinger was upset to find that the scientific basis for and conservatism of the interim ECCS criteria are now in doubt after he had been assured by AEC staff of their validity.[56]

Testimony by the reactor manufacturers during the summer produced no significant new evidence in support of the AEC reactor safety regulations. It thus developed that the industry's last chance to demolish the Intervenors' case against it would occur when Henry Kendall and Dan Ford took the stand in August 1972 to defend their 300-page written testimony.[57]

This portion of the hearing again opened with several days of legal dispute over Ford's qualifications to participate. This time the hearing board ruled that Ford could testify only on those portions of the UCS testimony that he had actually written. The Intervenors' attorney, Myron Cherry, argued that this worked an unnecessary hardship on Kendall, since Ford had attended the entire hearing so far while Kendall had not been able to do so; but Cherry was overruled. In any case, it was true that, although Ford had been responsibile for the gathering and preliminary evaluation of references, Kendall had done the actual technical analysis. Kendall was the physicist, and it was he who would have to defend his technical critique.

Henry Kendall, in his forties, is tall and rangy, obviously an outdoorsman. His manner is intense and his chiseled face, penetrating eyes, and sweptback dark blond hair give him a striking presence. Now a full professor of physics at MIT, Kendall has built a solid career as an experimental physicist while leading a remarkably active life. Inherited wealth has allowed him to follow his adventurous instincts. He has a considerable reputation as a mountain climber and mountain photographer, with a number of first ascents of 20,000-foot peaks in the Andes to his credit. He is also a skindiver and a private pilot. And finally, Kendall had been for a number of years a member of the elite "Jason" advisory group of Defense Department consultants, with which he had worked on both military and civilian problems.

During the legal maneuvering before Kendall took the stand, an industry lawyer gloated over what he claimed was the Intervenors' "gross lack of confidence in their testimony."[58] An AEC staff member even went so far as to invite a *New York Times* newsman to be present to report Kendall's expected demise. The jubilation in the reactor proponent ranks was premature, however, When the cross-examination actually began, Kendall fared rather well. Indeed, the cross-examination gave Kendall the opportunity to argue that his analyses were, if anything, overcautious: reactors might well be even *less* safe than his prepared testimony asserted. Finally, after surviving nearly two weeks of cross-examination with no serious setbacks (except for losing fifteen pounds!), Kendall faced his last challenger: Westinghouse.

Westinghouse is the largest American manufacturer of nuclear reactors, and its pressurized-water reactors have perhaps come under the heaviest attack for safety deficiencies. The Westinghouse team fared little better at beating Kendall down than its predecessors, however. Before long the Westinghouse lawyer, Barton Z. Cowan, was reduced to minor quibbling about the accuracy of quotations in the UCS testimony. Later Cowan announced that he would publicly discredit Kendall by quizzing him on his expertise with a list of questions from twenty-four technical disciplines, but Cowan never got beyond disciplines number 1 (hydraulics and fluid mechanics) and 2 (thermodynamics). Finally, Cowan asked Kendall and Ford rather sarcastically if it was not possible that the AEC staff was in a better position to evaluate reactor safety than the UCS. Ford responded by using the opportunity to express his misgivings about the conduct of the AEC staff in the ECCS controversy. He reminded the hearing board that the AEC task force that had devised the Interim Criteria had utilized data and analyses provided by the reactor manufacturers, while they had ignored (or never saw) independent analyses leading to different conclusions prepared at the AEC's own laboratories. In Ford's view, the staff's independence was further compromised by previously promulgated AEC positions. Ford concluded:

> The final reason which would seem to inhibit the Regulatory Staff's ability to perform an objective, credible, scientific assessment of the safety implications of reactor systems and the acceptability of emergency core cooling systems has to do with the divided loyalties that seem to be built into the Atomic Energy Commission by the legislation that set it up.
>
> The Atomic Energy Commission seems to have accepted the responsibility to promote nuclear power rather than to be the guardians of the public interest in nuclear affairs. And I think this dedication on the part of the Atomic Energy Commission is reflected in the criteria that we have been reviewing in this hearing and is reflected in the Regulatory Staff's inability to do the job they ought.[59]

After this the Westinghouse lawyer became even more sarcastic, asking: "Is there any area in ECCS where anybody knows more than you two fellows?"[60] This question gave Ford and Kendall an opportunity to explain eloquently what they saw to be their role in the hearings. They could equally well have been presenting a general argument for the necessity of public interest science:

FORD: Mr. Cowan, in terms of general knowledge of the field of emergency core cooling ... [we] would readily defer to the various people—thorough, competent, solid engineers—who have dedicated themselves to studying this field. . . .

> Now, I think that our function has been in part to assist these people in communicating with the Atomic Energy Commission by developing and cultivating this forum in which they can ... break through the various bureaucratic manacles that have prohibited them for so long from expressing what is a widely shared, deeply felt view in the nuclear community itself, among those persons intimately concerned with this area. . . .

KENDALL: [The] question here is a question of communication and of freedom to communicate, and not being able to speak freely. . . .

> These are qualified people in that Laboratory [i.e., Oak Ridge], and we all hold them in considerable respect. The difficulty is not that they do not know enough, it is that they are not heard. And the contribution that we believe that we can make is that we are in a position to be heard better than they. . . . We can speak relatively freely of institutional pressures, and say things that would otherwise have to be extracted with great difficulty from reluctant mouths.
>
> There is no question, Mr. Cowan, but that many of the people who have taken the stand here are professionals who have spent a good portion of their professional lives in this field and have available to them from memory many more facts with respect to emergency core cooling systems and with respect to nuclear reactor operation than I do.
>
> There is no question but what that facility is not the critical and important facility for the kinds of things that are under discussion in this hearing, because what is called for here is a question of judgment, first, and second, a position from which one can speak freely.[61]

The Hearings Conclude

The rebuttal phase of the AEC hearings on reactor emergency cooling systems was followed by a second round of testimony in late 1972 that produced few surprises, and then by the submission of closing statements by all parties in early 1973. In their closing statements, all of the reactor manufacturers except General Electric contended that the AEC's ECCS Interim Criteria were too conservative and should be weakened, while General Electric was willing to accept them as they stood but was quite certain that they should not be made any more stringent.[62] On the other side, Kendall and Ford argued that, in view, of the inadequate experimental understanding of the actual behavior of emergency cooling systems—a deficiency that had been brought out by their own testimony and by that of AEC reactor safety experts—the Interim Criteria were without justification and the AEC had no basis for licensing water-cooled reactors.

The AEC regulatory staff, as participants in the hearing, also submitted a closing statement. The recommendation which it contained displeased both the reactor manufacturers and the Intervenors. The regulatory staff proposed new reactor licensing criteria that were slightly more conservative and filled in some of the gaps which had been exposed in the Interim Criteria. The regulatory staff speculated that some reactors might even be "derated"—forced to operate below their full power levels—by as much as 20 percent until their emergency cooling systems could be upgraded to meet the new criteria. Others doubted that any such derating would actually result.[63] Kendall termed the changes largely "cosmetic" and emphasized once again that the fundamental problem lay, not with the details of the criteria themselves, but instead with the lack of the basic knowledge required to assure that, in the event of a loss-of-coolant accident in a major nuclear reactor, its emergency cooling system would be able to prevent a catastrophic release of radioactivity into the environment.[64]

The final decision on whether and how much to modify the ECCS Interim Criteria was issued by the AEC Commissioners themselves more than a year later on December 28, 1973. The Commissioners essentially adopted the criteria proposed by the regulatory staff.[65]

AEC Licenses Reactors Anyway

The national hearings on the ECCS Interim Criteria were orginally convened because the AEC had failed, in December 1971, to convince the local hearing board on the Indian Point 2 reactor of the adequacy of the Interim Criteria. The issue had first been publicly articulated in reports by the Union of Concerned Scientists, and it had been forcefully presented in the Indian Point 2 hearings in the course of Dan Ford's appearance there as a technical interrogator. The Indian Point 2 board's action automatically set a precedent for all other reactor licensing hearings, raising the possibility that the entire licensing program might grind to a halt, leaving billions of dollars' worth of completed nuclear power plants idle. It therefore seemed reasonable for the AEC to propose national hearings on the issue so that the same ground would not have to be worked over in each local hearing, and the local intervenors agreed to cooperate.

But then, in autumn 1972, with the national hearings still in midstream, the AEC suddenly instructed its local hearing boards to disregard the emergency core-cooling issue and proceed with the licensing of seventeen new nuclear power plants. The AEC contended that these plants were badly needed and that they were safe enough. The Consolidated National Intervenors felt betrayed. For a year and a half they had worked within the AEC's administrative procedures. And now, before the final judgment was in, they saw the AEC committing itself to the design standards of current nuclear power plants. Henry Kendall concluded that the outcome of the national hearings was a foregone conclusion, and that the hearings had been used by the AEC mainly as a device to remove the troublesome safety question from the licensing hearings on individual nuclear power plants during the crucial period when nuclear power was finally coming "on line" on a large scale. Shaw's sabotage of the AEC's own safety program during this period provided additional basis for this cynical view.

Time was indeed running out for the Intervenors. While the local hearings on reactor operating licenses and the national hearings on the reactor ECCS Interim Criteria ground on, the hard-pressed electric utility companies continued to order new nuclear power plants. In 1972 the capacity of the nuclear reactors already operating, under construction, or on order in the United States amounted to some 127 million kilowatts, about 40 percent of the total electric power generating capacity in existence in 1970.[66] By 1976 or so, when many more of these nuclear plants will be in operation, shutting them down would be so disruptive that even a major catastrophe might not bring that about.

The AEC doubtless should have followed the advice of its own Advisory Committee on Reactor Safeguards in 1966, when construction began on the present generation of billion watt reactors, and should have pressed a serious program of research on reactor safety. It may still not be too late for a crash program of reactor safety research. The emergency core cooling problem is basically an engineering problem, difficult but probably not insoluble if the reactor industry and the AEC give it sufficiently high priority. It is encouraging that the reactor manufacturers have been redesigning reactor cores for operation at lower power density, for greater controllability in the event of an accident. Westinghouse is reportedly also designing a new improved emergency core cooling system.[67]

Beyond the Hearings

Despite the frustrations of the ECCS hearings, they gave the Union of Concerned Scientists an opportunity to get the facts out into the open, including the AEC's own reactor experts' data and opinions. On the basis of that record, Ford and Kendall decided in autumn 1972 to build a fire of public concern under the AEC and Congress's Joint Committee on Atomic Energy. Their efforts were greatly aided by a series of in-depth articles in *Science* magazine by reporter Robert Gillette,[68] which were followed by regular coverage of the subject in the *New York Times* and other leading papers. And on May 31, 1973, ABC television screened an hour-long documentary on nuclear reactor safety featuring interviews with Ford and Kendall as well as with Milton Shaw and several AEC Commissioners. Ralph Nader had become interested in reactor safety in late 1971, but hesitated to associate himself with the Consolidated National Intervenors until he had studied the issue in detail. By January 1973, on the basis of the ECCS hearings record and the personal presentations of Kendall and Ford, Nader decided to join forces with them. Thereafter he repeatedly endorsed UCS positions and attacked the AEC in press conferences, speeches, and television appearances.[69] The UCS also involved itself in a major debate over nuclear power which has developed in California, where Ford and Kendall have testified on reactor safety before the state's Public Utilities Commission and the state legislature.

AEC Reorganization

In 1973, partly as a consequence of changing leadership and partly in response to the political and legal pressures generated by the Intervenors and their allies, the AEC made some moves to reorganize its efforts on reactor safety. In May 1972 Senator Howard Baker (R.-Tenn.), in whose home state Oak Ridge National Laboratory is located, had tried unsuccessfully to convince his colleagues on the Joint Committee on Atomic Energy that reactor safety research should be separated from Milton Shaw's AEC Division of Reactor Development and Technology.[70] Both then and on several later occasions, AEC Chairman James Schlesinger joined with senior members of the Joint Committee in staunch support of Shaw. But in January 1973, Schlesinger was moved by President Nixon to the directorship of the Central Intelligence Agency, and he was succeeded in the AEC chairmanship by Dixy Lee Ray, a marine biologist from Seattle.

Dr. Ray is a somewhat unusual woman who lives with two dogs in a mobile home in suburban Maryland. At first she was not taken very seriously either within or outside the AEC. But after biding her time for a few months, in mid-May 1973 she acted swiftly and decisively to force through a substantial

AEC reorganization along the lines that had been proposed a year earlier by Senator Baker. With the help of the new members on the Joint Committee on Atomic Energy, she was also able to secure the Joint Committee's approval. Although Dr. Ray insisted that the reorganization was not meant as a personal attack on Shaw, his office was stripped of all responsibility for conventional reactor safety research and left to concentrate on developing the breeder reactor. Shaw himself was said to be "absolutely furious" and threatened to quit the AEC, according to the nuclear industry trade press, while AEC safety researchers were reported to be "dancing in the streets" at the National Reactor Testing Station in Idaho. Dr. Ray demanded and received Shaw's resignation a few weeks later.[71]

In announcing the AEC reorganization, Dr. Ray said that it would provide for

greater emphasis and effectiveness in our safety research programs . . . [and give] new directions and a renewed dedication to safety research which will help speed resolution of the still unanswered questions.[72]

Asked whether she expected substantive changes under the new director of reactor safety research, Dr. Herbert Kouts, a former chairman of the Advisory Committee on Research Safeguards, her response was "Good heavens, I would hope so."[73]

Conclusions

The reactor safety issue is still far from settled. If reactors which the electric utility companies are building all over the country prove to be unsafe, the nation may have to learn to live with periodic radiological disasters in addition to the usual fare of hurricanes, earthquakes, floods, and the steady toll of automobile accidents. Or perhaps serious reactor accidents will be exceedingly infrequent—perhaps not even one by the end of this century. One hopes that the latter eventuality is more likely, but one would like to be in a position to say so with greater assurance.

No matter how the reactor safety issue is finally resolved, three lasting conclusions can already be drawn. First, with respect to the AEC: even if the conflict-of-interest issue had never been raised before, the present reactor safety controversy illustrates convincingly the unacceptable situation created by lodging responsibility for promoting and regulating nuclear power in one and the same agency. Since the subversion of the AEC's regulatory function has been encouraged or condoned at the level of the AEC Commissioners and the Congressional Joint Committee on Atomic Energy, the recent reorganization at a lower level has not dealt with this central problem. It is furthermore intolerable that the hearings on reactor safety had to be conducted before an AEC-appointed board, with official AEC witnesses defending an AEC-approved position and unofficial AEC witnesses subjected to threats of losing their jobs, with AEC documents and advisors not subject to subpoena, and with the final decision in the hands of the AEC Commissioners. In any fair and impartial hearing, the government agency charged with regulating reactors and protecting the public interest would have itself prepared the testimony that the Consolidated National Intervenors were forced to draw "from reluctant mouths," to use Kendall's phrase. Indeed, if an independent agency were charged with nuclear safety, it seems probable that the problems of emergency core cooling would have been dealt with much earlier and in a more adequate manner.

A second conclusion is that one does not have to be the world's greatest expert to challenge even so mighty and technically esoteric an agency as the AEC. Great effort and dedication are required; most important of all is good judgment, self-confidence, and independence of mind. The true reactor safety experts at the AEC laboratories responded to the dedication and competence of Kendall and Ford and undertook to educate them and cooperate with them.

Finally, it is difficult to avoid being struck by the multiple failures of our scientific institutions that this reactor safety controversy has revealed. The supposedly independent AEC Advisory Committee on Reactor Safeguards had been quietly warning the Commission of serious deficiencies in reactor safety research ever since 1966. But these warnings fell on deaf ears, and the urgency of the need for additional information and safer reactor designs did not become apparent even to the larger nuclear science community until the ECCS hearings began. And even these hearings did not result from a demand by nuclear reactor engineers for an airing of all the relevant information. They came about because of the willingness of one economist and a few physicists to look into important issues far from their normal areas of expertise and to interject these issues into a reactor licensing controversy initiated by environmentalists. The costs of preparing and presenting the technical arguments have been borne by environmental groups like the Sierra Club and the Audubon Society; Kendall and Ford themselves have personally raised a substantial fraction of the $200,000 which they have spent thus far. Their professional colleagues and scientific organizations have not been among the major contributors.

Kendall and Ford are among the pioneers in public interest science. Their achievements will hopefully inspire others.

NOTES

1. U.S., Atomic Energy Commission, *In the Matter of Acceptance Criteria for Emergency Core Cooling Systems for Light Water Nuclear Power Reactors,* Docket No. R.M. 50-1 (Hereafter referred to as *ECCS Rule-Making Hearing.*) transcript, p. 19299.
2. *Federal Register,* April 23, 1971, p. 7696.
3. *Boston Globe*, April 1971.

4. The March 4 lectures are reprinted along with a brief history of the March 4 movement in Jonathan Allen, ed., *March 4* (Cambridge: MIT Press, 1970). See also Dorothy Nelkin, *The University and Military Research: Moral Politics at M.I.T.* (Ithaca, N.Y.: Cornell University Press, 1972).

5. See, e.g., Ralph E. Lapp, "The Nuclear Power Controversy: Safety", *New Republic*, January 23, 1971, p. 18.

6. For a pungent popular presentation of the views of these authors, see their joint efforts *"Population Control" Through Nuclear Pollution* (Chicago: Nelson Hall, 1970) and *Poisoned Power* (Emmaus, Penn.: Rodale Press, 1971). Although these books have a slightly hysterical tone, their facts appear to be basically correct. (See *The Effects on Populations of Exposures to Low Levels of Ionizing Radiation* [Washington D.C.: National Academy of Sciences, November 1972].) The books also give an almost unique inside view of some of the problems which have arisen from the AEC's dual role of promoter and regulator of atomic energy.

7. *New York Times,* June 8, 1971, p.1.

8. See, e.g., Ralph Lapp, "The Nuclear Power Controversy: Power and Hot Water", *New Republic,* January 6, 1971, p. 20.

9. *New York Times,* March 17, 1971, p. 29.

10. The tests applied most directly to "pressurized-water reactors," which are built by Westinghouse, Babcock and Wilcox and Combustion Engineering. The other common variety of modern power reactor is General Electric's "boiling-water reactor," which also requires an elaborate emergency cooling system. A third, less common design is the Gulf General Atomic "high-temperature gas-cooled reactor."

11. The AEC supports a number of major laboratories. The experts on the safety problems of conventional power reactors were in 1970 located mostly at the Oak Ridge National Laboratory in Tennessee and the National Reactor Testing Station (NRTS) in Idaho. Both have staffs of several thousand employees, but Oak Ridge is one of the AEC's "multi-purpose laboratories" with programs in many areas of physical and biological research and engineering while the NRTS is more "mission oriented." Oak Ridge is operated for the AEC by Union Carbide and the NRTS by Aerojet-General. The Lawrence Livermore Laboratory, mentioned previously, is operated for the AEC by the University of California and has as its primary mission the refinement of nuclear weapons.

12. For an illustrated description see Tom Alexander, "The Big Blowup Over Nuclear Blowdowns," *Fortune,* May 1973, p. 216. The illustrations include color photos of fuel rods swollen and contorted in tests at Oak Ridge simulating conditions in an overheated reactor core.

13. This "once through" cooling water is not run through the reactor core. Instead, it is used to recondense the steam which has delivered the energy of the reactor to the turbines which drive the electric generators. The "primary cooling water" which passes through the reactor inevitably becomes contaminated with radioactivity and must be recycled.

14. U.S. Atomic Energy Commission, "Theoretical Possibilities and Consequences of Major Accidents in Large Nuclear Power Plants," AEC Report No. WASH-740 (Washington, D.C.: AEC, 1957).

15. Ian A. Forbes, Daniel F. Ford, Henry W. Kendall, and James J. MacKenzie, *Nuclear Reactor Safety: An Evaluation of New Evidence* (Cambridge, Mass.: UCS, July 1971). (The address of the UCS is P.O. Box 289, MIT Branch Station, Cambridge, Mass. 02139.) The report is reprinted in U.S. Congress, Senate, Committee on Interior and Insular Affairs, *Calvert Cliffs Court Decision,* 92nd Congress, 1st Session, November 3, 1971, Part 2, pp. 295-305; the passages quoted appear on p. 299. This UCS report is also reprinted in part in *Nuclear News,* September 1971, p. 32. (*Nuclear News* is the monthly organ of the American Nuclear Society.)

16. *Ibid.* (*Calvert Cliffs Decision*), p. 300.

17. The delays in the AEC's nuclear safety program and the resulting controversy within the AEC establishment are discussed in a series of articles by Robert Gillette titled "Nuclear

Safety" and subtitled "The Roots of Dissent," *Science* 177 (1972): 771; "The Years of Delay," p. 867; "Critics Charge Conflicts of Interest," p. 970; and "Barriers to Communication," p. 1030.

18. Forbes et al., *Nuclear Reactor Safety*, note 15.

19. The text of the Seaborg letter is reprinted in *Nucleonics Week,* May 13, 1971, p. 6. (*Nucleonics Week* is a McGraw-Hill trade newsletter.)

20. Quoted in Robert Gillette, "Nuclear Reactor Safety: A Skeleton at the Feast?" *Science* 172 (1971):918. Gillette notes that the Joint Committee had already received a secret AEC briefing on the emergency cooling system issue on May 3, 1971.

21. "Criteria for ECCS for Light Water Reactors," *Federal Register* June 29, 1971, p. 125.

22. R. Lorenz, D. Hobson, and G. Parker, *Final Report on the First Fuel Rod Failure Transient Test of a Zircaloy-Clad Fuel Rod Cluster in TREAT,* Report No. ORNL-4635 (Oak Ridge, Tenn: Oak Ridge National Laboratory, March 1971).

23. Quoted in Robert Gillette, "Nuclear Safety: Critics Charge Conflict of Interest", *Science,* 177 (1972):970.

24. Daniel F. Ford, Henry W. Kendall, and James J. MacKenzie, *A Critique of the New A.E.C. Design Criteria for Reactor Safety Systems* (Cambridge, Mass: UCS, October 1971). Reprinted in *Calvert Cliffs Court Decision* pp. 306-319. (Also reprinted in part in *Nuclear News*, January 1972, p. 28.) The contents of this and the UCS July report (Forbes et al., *Nuclear Reactor Safety*) are summarized in Ian A. Forbes, Daniel F. Ford, Henry W. Kendall, and James J. MacKenzie, "Cooling Water," *Environment,* January-February 1972, p. 40.

25. Letter from Donald B. Trauger, Associate Director, Oak Ridge National Laboratory, to Milton Shaw, Director, Division of Reactor Development and Technology, AEC, November 24, 1971. Entered as exhibit F, testimony of Intervenors' counsel, Myron Cherry, *ECCS Rule-Making Hearing,* March 17, 1972. The letter is reprinted in part in *Congressional Record* 118 (1972):10450.

26. *Nucleonics Week,* April 20, 1972, p. 1.

27. Quoted in *Nuclear Industry,* February 1972, pp. 11-20. (*Nuclear Industry* is published by the Atomic Industrial Forum, a nuclear industry association.)

28. Richard D. Lyons, "Nuclear Experts Share Doubts on Power Plant Safety," *New York Times,* March 12, 1972.

29. Morris Rosen and Robert J. Colmar, "Comments and Recommendations to the REG ECCS Task Force," memorandum of June 1, 1971. Exhibit No. 715 of the ECCS Hearing record; reprinted in the *Congressional Record* 118 (1972):9303, 9304.

30. Prepared testimony of Dr. Morris Rosen, Technical Advisor to the Director, Division of Reactor Licensing, *ECCS Rule-Making Hearing,* Exhibit 1043, p. 34.

31. *Ibid.,* p. 40.

32. Quoted in Robert Gillette, "Nuclear Reactor Safety: At the AEC the Way of the Dissenter is Hard," *Science* 176 (1972):498.

33. Quoted in *Nucleonics Week,* March 2, 1972, p. 11.

34. Quoted in Gillette, "Nuclear Reactor Safety: At the AEC the Way of the Dissenter is Hard," p. 498.

35. Quoted in *Nucleonics Week,* March 2, 1972, p. 1.

36. *ECCS Rule-Making Hearing*, transcript, pp. 2114-2120.

37. Quoted in Robert Gillette, "Nuclear Reactor Safety: At the AEC the Way of the Dissenter is Hard," *Science,* 176 (1972):496.

38. Quoted in Daniel F. Ford and Henry W. Kendall, "Nuclear Safety", *Environment,* September 1972, p. 48, footnote 6.

39. Quoted in *Nucleonics Week,* March 16, 1972, p. 2.

40. Letter from William B. Cottrell et al. to L. Manning Muntzing, Director of AEC Division of Regulation, December 6, 1971. Submitted as Exhibit No. 1020 at the *ECCS Rule-Making Hearing*; reprinted in part in the *Congressional Record* 118 (1972):10449.

41. Gillette, "Nuclear Reactor Safety: At the AEC the Way of the Dissenter is Hard," p. 498.

42. Letter from Dr. Alvin A. Weinberg to Dr. James R. Schlesinger, February 9, 1972. Exhibit No. 1027 at the *ECCS Rule-Making Hearing*; reprinted in part in the *Congressional Record* 118 (1972):10449.

43. Gillette, "Nuclear Safety: Critics Charge Conflicts of Interest," p. 972.

44. *ECCS Rule-Making Hearing,* transcript, p. 7289.

45. *Ibid.*, pp. 7592-7593.

46. *Nucleonics Week,* September 7, 1972, p. 8.

47. Gillette, "Nuclear Safety: The Roots of Dissent," p. 773.

48. Member of Weinberg's staff, December 1971, personal communication.

49. See e.g. *ECCS Rule-Making Hearings,* transcript, p. 724. See also *Nuclear Industry,* April 1972, p. 9.

50. *ECCS Rule-Making Hearings,* transcript, pp. 7382-7383.

51. *Ibid.*, p. 7387.

52. *Ibid.*, pp. 7434-7435.

53. *Nucleonics Week,* April 13, 1972, pp. 4-5.

53. *Ibid.*

54. H. Peter Metzger, *The Atomic Establishment* (New York: Simon and Schuster, 1972), p. 244. Metzger's book is a hard-hitting presentation of the history and consequences of the insulation from outside criticism of the AEC and the Joint Committee on Atomic Energy.

55. Quoted in *Nucleonics Week,* February 17, 1972, p. 1.

56. *Nucleonics Week,* April 20, 1972, p. 1.

57. Kendall and Ford were assisted in preparing this testimony by two Union of Concerned Scientist Colleagues, Professors James A. Dawson of the Harvard School of Public Health and James V. Fay of the MIT department of mechanical engineering. Fay wrote one of the chapters.

58. Quoted in *Nucleonics Week,* August 3, 1972, p. 3.

59. *ECCS Rule-Making Hearing,* transcript, pp. 19266-19267.

60. *Ibid.*, p. 19267.

61. *Ibid.*, pp. 19267-19278.

62. *Nuclear Industry,* March 1973, p. 40.

63. Robert Gillette, "Reactor Safety: AEC Concedes Some Points to Its Critics," *Science* 178 (1972):482.

64. *Not Man Apart,* December 1972, p. 1. (*Not Man Apart* is the monthly magazine of Friends of the Earth.) Kendall and Ford, private communication.

65. U.S. Atomic Energy Commission, *Acceptance Criteria for Emergency Core Cooling Systems, Light-Water-Cooled Nuclear Power Reactors, Opinion of the Commission,* Docket No. R.M. 50-1.

66. *Nuclear News,* 1972 International Conference Program, p. 107.

67. See, e.g., *Nuclear Industry,* July 1972, p. 27; *ibid.,* October 1972, p. 48; *Nucleonics Week,* October 26, 1972, p. 2. The Electric Power Research Institute, which is funded by the electric utility industry, has also undertaken the funding of its own program of research on nuclear reactor safety. See, e.g., John Walsh, "Electric Power Research Institute: A New Formula for Industrial R & D," *Science* 182 (1973):263.

68. Gillette, "Reactor Safety," note 17.

69. See, e.g., *New York Times,* January 4, 1973, p. 18.

70. *Nucleonics Week,* May 4, 1972, p. 4. Senator Baker is an old friend of Alvin Weinberg, then director of the Oak Ridge National Laboratory. Baker admitted consulting with people at Oak Ridge before making his proposal, but he denied that it was Weinberg's idea.

71. *Nucleonics Week,* issues of May 17, 24, 31, 1973. See also *Science* 180 (1973):935.

72. *New York Times,* May 22, 1973, p. 17.

73. *Nucleonics Week,* May 31, 1973, p. 3.

PART V

Public Interest Science

CHAPTER 16

When Outsiders Can Be Effective

The examples of public interest science activities described in the preceding chapters are extremely varied. They involved the courts, Congress, federal agencies, and state and local governments. Some fights were over in a matter of months while in other cases the battle wore on for many years. But there is a unifying theme in all these cases: they all involved scientists and citizen groups trying to change government policies by presenting their criticisms and recommendations as effectively as they could in the most favorable forum that they could find. In this chapter we try to abstract some of the lessons that these case studies have to offer about when and how outsiders can be effective.

Easy Fights

Sometimes, when there are no great vested interests involved, it is not difficult to change government policy. The practice in question may be simply a matter of thoughtlessness, and thus when it becomes a political embarrassment the agency responsible may move quickly to rectify the situation. This was what happened twice, for example, after the Colorado Committee for Environmental Information disclosed that the Army was storing nerve-gas bombs under the approach path to Denver's airport. The plan to send twenty-odd trainloads of chemical weaponry rumbling through cities across the country for eventual dumping in the Atlantic Ocean off New Jersey was the Army's idea of an easy way out of its embarrassment. Then, when the proposed rail shipment was revealed and provoked general outrage, the Army quickly switched signals and agreed to follow a National Academy of Sciences panel's recommendation to detoxify the obsolete gas in place. Then, when the public relaxed, thinking that the issue was settled, the Army relaxed, too, and the detoxification program wallowed in technical difficulties—again the easy way out. Most recently, in summer 1973, the Army changed course once again and agreed to begin detoxifying the nerve gas bombs immediately in response to renewed pressure from the citizens of Denver.

In contrast to this case, where the resistance to changing the criticized practices was lackadaisical, the critics of the federal ABM, SST, and pesticide-regulation programs encountered the most bitter opposition. Here they were attacking policies that involved billions of dollars. As a consequence, the battles were rough and prolonged and required the active involvement of large numbers of citizens in addition to scientists.

Hard Fights

In hard-fought cases the success of the outsiders depends upon a number of factors, including the timeliness of the issue, whether it poses a personal and obvious danger to individual members of the middle or upper class public, the existence of an appropriate forum, the special visibility of certain issues in particular localities, and the credibility of the public interest scientists themselves.

TIMELINESS

The influence of scientist-advocates has often depended upon the timeliness of an issue. Thus, after Bo Lundberg and others had denounced the SST for years with little apparent effect, the new environmental movement in the late 1960s came to see the SST as a symbol of all that is destructive to the environment—and found it a ready-made issue complete with documentation.

Similarly, in the case of defoliation in Vietnam, the protests of a few biologists and ecologists went unheard for several years until the American public became disgusted with the entire United States Indochina policy. Only then was the American Association for the Advancement of Science willing to take the step of funding the Herbicide Assessment Commission's expedition to Vietnam. And when the HAC returned, it found an audience willing to hear its distressing findings.

Finally, in yet another case, the ABM became a popular issue in part because the dissenting scientists took their case to the public at a time when the insatiable appetite of the military-industrial complex was becoming a matter of popular concern. Cost overruns and the failure of new weapons systems to meet their performance specifications, along with the well-advertised mismatch between the Army's words and deeds in Indochina, had eroded the public's usual willingness to provide the Pentagon with a blank check.

PERSONAL AND OBVIOUS DANGER

One feature that all these public campaigns have in common is that their success depended on large numbers of people being able to see the technologies under attack as a potential threat to themselves personally. Consider the ABM debate. For years, scientists and strategists had argued the relative merits of various nuclear weapons systems, but the average educated person generally ignored the debate. The matter of strategic weapons was cloaked in technical jargon and military secrecy, and their destructiveness, while undeniably enormous, seemed remote and impersonal. But when the Johnson administration decided to place Spartan antimissile missiles armed with multimegaton hydrogen bombs in the suburbs of some of the nation's largest metropolitan areas and it was pointed out that an accidental explosion would have very obvious and personal consequences for large numbers of people, the reaction of suburbanites was direct and politically potent. The Nixon administration was forced to ban the ABMs to faraway North Dakota where they were given a new mission: guarding missiles instead of people.

William Shurcliff made the issue of the SST similarly direct and personal. He pointed out that the sonic boom from each transcontinental supersonic transport flight would annoy everyone in a path some fifty miles wide stretching from coast to coast. The popular response finally forced the government to promise that SSTs would not be allowed to fly over the United States.

In both of these cases the government moved to accommodate the public's concerns in an attempt to save the programs. But in neither case did the political reexamination of these programs stop at this point. The "bombs in the backyard" and the sonic-boom issues served to make the ABM and SST programs respectively visible to Congress and the nation, and they remained front-page news for some time thereafter. An overall reexamination of these programs followed which ultimately led to their demise.

AN APPROPRIATE FORUM

Not every issue conjures up in the minds of the public the fear of a mushroom cloud, of a picture window broken to shards by a sonic boom, or of some other such dramatic event. And the public does not and cannot respond effectively to all of the important issues which are presented to it directly. Fortunately society offers other, less political forums in which some of these issues can be dealt with—in particular, judicial and administrative hearings.

The case of DDT is a good example. The steadily accumulating level of DDT in the biosphere worried nature lovers, who saw the damage already being suffered by wildlife. But the danger to man, even when articulated by Rachel Carson in her powerful book *Silent Spring,* was not clear to the general public. Opposition to the use of DDT was largely limited to "birdwatchers" and scientists.

Enter the Environmental Defense Fund. Bypassing the politically entrenched pro-DDT forces in the Agriculture Department and Congress, the EDF sued in the courts to block unnecessary use of DDT on Long Island and then in western Michigan. Although the courts ultimately refused to assume jurisdiction, the evidence presented—of DDT's lack of efficacy, ecological harmfulness, and likely carcinogenicity—convinced local officials to stop using it anyway. The EDF kept up the legal pressure, and ultimately, after several rounds in the District of Columbia's Court of Appeals, they forced the federal government to ban DDT altogether.

The courts. The role which the courts played in this case is fairly typical. They did not themselves decide the merits. Rather, they considered whether the responsible government agencies had taken adequate account of the hazards involved in the use of DDT. This allowed the opponents to put the case against DDT into the record, after which the court would as often as not agree with them that the government agency had not done its job properly and would order the agency to try again.

It might seem to be a futile gesture to return an issue in this way to an agency which is politically committed to a particular policy, but in practice this has not been so. A court decision that an agency has not done its job properly can be a tremendous blow to that agency's credibility and can, for example, encourage a previously reluctant state government to make up its own mind. This seems to have been the effect, in a number of states, of the decisions of the U.S. Court of Appeals on the suits brought by the Environmental Defense Fund against the U.S. Department of Agriculture and the Environmental Protection Agency.

Recent developments in the law, particularly the National Environmental Policy Act of 1969 with its requirement of comprehensive "environmental impact statements" on federal actions significantly affecting the quality of the human environment, have greatly increased the jurisdiction of the courts on environmental issues. Thus, the courts can provide an alternative forum for scientist action in issues that for reasons of technical complexity, lack of public interest, or political entrenchment of vested interests are unsuitable for a public campaign aimed at Congress. A small number of scientists can have a tremendous impact in the courts if they have a good case and are able to call upon their colleagues for expert testimony. Only a half-dozen scientists organized the entire Environmental Defense Fund campaign against DDT, but their efforts were supported by the testimony of more than a hundred expert witnesses.

Administrative hearings. The hearing on DDT held by the Wisconsin Department of Natural Resources illustrates another forum for public interest science: the administrative hearing. Other examples are the protracted administrative hearings on the effectiveness of emergency core-cooling systems begun in January 1972 by the Atomic Energy Commission and the administrative hearings on DDT held by the Environmental Protection Agency in 1971 and 1972. Each of these hearings allowed critics to lay out at least some of the issues for the record, irrespective of the sympathies of the sponsoring agency. Even when the finding of the hearing examiner was adverse to the critics' cause—as occurred with DDT—other interested groups were able to draw their own conclusions. Thus, the Administrator of the Environmental Protection Agency disagreed with the hearing examiner and found the case against DDT persuasive. And on the nuclear safety issue some influential segments of the media were shocked by what the hearing record showed of the internal workings of the Atomic Energy Commission, and the AEC, under a new chairman, did some house-cleaning as a result.

LOCAL DEBATES

A number of our case studies involve local controversies which grew into national debates. This is what happened after the Environmental Defense Fund had put its show on the road for two years. The issue had developed to the point where it could play to the audience in Washington.

The EDF has applied this technique to other issues. A majority of its hundred-odd current legal actions concern local rather than national issues: saving an unspoiled river, stopping an industrial polluter, or suing for changes in state electricity rate structures. But the EDF Board of Trustees tries to choose its cases so that they will establish precedents applicable elsewhere.

As another example, the national controversy over the safety of nuclear reactors began when the issue was introduced into the licensing hearings for particular reactors. Similarly, the local controversy in Colorado over the storage of nerve gas at the Rocky Mountain Arsenal served to dramatize the national debate over U.S. chemical and biological warfare policies. And finally, the national debate over the Sentinel and Safeguard antiballistic missile systems developed out of local campaigns against particular ABM sites in the Seattle, Chicago, and Boston areas.

One of the advantages of working locally is that a few scientists with a good case can not only get excellent local news coverage, but can also personally meet with and have an opportunity to convince local decision makers: mayors, town councilmen, and other municipal and state officials.

CREDIBILITY

From the first moment that he raises a criticism of an accepted government policy, the public interest scientist is confronted with the question: "Why should we take your word over that of government officials—who, after all, have the best experts at their disposal? How do we know that you're not some kind of kook?" Different groups have used different methods to combat this credibility problem:

- • • Rachel Carson published a compelling and well-documented book on the misuse of pesticides. It didn't convince everyone, but it made certain that her arguments received a hearing.
- • • The Herbicide Assessment Commission was sponsored by a recognized scientific institution, the American Association for the Advancement of Science.
- • • The Union of Concerned Scientists got excellent mileage out of quoting AEC-sponsored studies whose conclusions contradicted the official AEC line on reactor safety.

Yet another technique for dealing with the credibility problem is to shift the question to the opponent, as did the Colorado Committee for Environmental Information in the controversies over plutonium pollution and natural-gas stimulation. In each of these debates, the CCEI publicly challenged the responsible government agency to establish the basis for its assertions that the public was not at risk. The Colorado group followed up its challenge with a specific list of technical questions, the answers to which would make possible an independent determination of public safety.

Public interest science is of course not without its "exaggerators." But there are surprisingly few. A scientist's reputation is his most precious possession, and the scientist who misrepresents the truth or makes unsound technical judgments calls down upon himself the censure of his colleagues. Furthermore, technical arguments presented in public can be rebutted in public, in the usual self-correcting manner of scientific discourse.

It is important that high standards be maintained by public interest scientists. They have enough difficulties as it is getting a hearing for important issues without adding a "credibility gap" to their problems. Obviously, the proper ethics for outsider science advising deserves discussion within the scientific community no less than do the ethics of insiders. Since in Chapter 9 we proposed two guidelines for federal executive-branch science advisors, perhaps we should add at this point two for public interest scientists:

1. A specialist should not use his authority to lend support to a political position without stating the technical grounds for his opinion.
2. The standards of accuracy to which a scientist adheres in public statements should be no lower than those he strives to attain in his scientific work.

It is also necessary for the scientist to maintain a sense of perspective; it is all too easy to exaggerate the significance of an issue with which one is concerned to the point where attention is distracted from what may be an even more important issue.

The News Media

As must be clear from our case studies, the news media's treatment of technological controversies determines to a large extent the effectiveness of public interest science efforts. Unfortunately, the media have not exactly covered themselves with glory in their reporting of technological controversies.

WHY THE MEDIA DON'T LIKE TO GET STORIES FROM INDEPENDENT SCIENTISTS

Few mass-media reporters have sufficient technical background or are allowed by their editors to specialize enough to become familiar with the issues in a particular area of technology. As a result, most of them do not have confidence in their ability even to separate crackpots from competent scientists and engineers—and checking around would take more time than they are given for a story.

The few trained science reporters generally stay away from the more controversial areas of applied science and instead undertake to educate and entertain their readers with the latest nuggets from the research laboratories. Their stories range in style and substance from the "gee whiz" variety on death rays, test-tube babies, or the latest from the current space extravaganza to Walter Sullivan's excellent (albeit somewhat breathless) reports in the *New York Times* on the latest discoveries in astronomy or elementary-particle physics. This emphasis may partly result from scientists' reluctance to discuss with the press such issues as the side effects of cyclamates or the safety of nuclear reactors. Many scientists evidently regard such controversies as the dirty linen of science. Finally, editors usually have plenty of "real" news that will be of obvious interest to their readership—official corruption, rapes, inflation, and so forth—and a story on the possible effects on the arms race of a new strategic weapons system is less likely to "sell." If the story reports that some little-known self-appointed guardian of the public interest has attacked one of the nation's largest advertisers, that is an added incentive not to use it.

OFFICIAL SOURCES

A lot of what's happening in the country today, a lot of what's most vital in peoples' lives, isn't institutionalized, so there's no official spokesman for it. If you stick to covering the official sources, inevitably you miss a lot of important things that are going on elsewhere. So, for instance, the press largely missed one of the great migrations of human history, the migration of black people out of the South and into the cities, until Watts blew up in 1965. And until Ralph Nader made something sensational out of it, we missed the rise in consumer consciousness; now, ironically, we've made something of an official source out of Ralph Nader. It's the way we like to work.[1]

–Tom Wicker *(New York Times*
editor and columnist)

Perhaps the biggest problem in trying to alert the press to important technological issues is that most reporters have too little time and know too few sources of information to do serious investigative reporting. As a result, reporters tend to rely largely, if not exclusively, upon "official sources" for such news—mainly government officials and corporation spokesmen. All too rare is the reporter who checks out a self-serving government report—even to the extent that Christopher Lydon did when the Department of Transportation announced that its technical advisors had concluded that the SST could be made as quiet as conventional jets. By the simple expedient of telephoning the chairman of the advisory committee, Lydon found that this noise reduction was to be achieved by the use of noise suppressors whose weight was nearly equal to the plane's entire payload.[2]

Ironically, one welcome by-product of both the Indochina war and the Watergate scandal has been the inculcation in the press of a wary and skeptical attitude toward official sources of news. But it is not enough merely to be critical in reporting official statements: as Tom Wicker points out in the passage quoted above, it is also necessary to look at issues that officials are not even talking about.[3] And the indispensible role of the independent scientist-activists— the Rachel Carsons, William Shurcliffs, and Matthew Meselsons—is to bring such issues to our attention.

"OBJECTIVITY"

Probably the greatest difficulty confronting a scientist with a story that he wants to get into the press is the very definition of "news." His story may concern the air pollution from a particular industrial plant or the desirability of citizen intervention in the licensing of a new nuclear reactor, but as long as the headline is of the form "Scientist Says Such and Such," the story is likely to run on page 25, if at all. On the other hand, if the President blames the energy crisis on the environmentalists, the event itself is considered newsworthy. In other words, a problem must be associated with an "event" in order to be considered reportable: every story must have a "news peg." Most reporters and editors seem to feel that "objectivity" requires only that they report such "news"; "muckraking" seems to them too much like trying to manufacture news.

But scientists are temperamentally indisposed toward staging demonstrations or other pseudoevents in order to get news coverage. The most that they will usually do is release a report. Such a report, if it is covered at all, is at best the sensation of a day; if it is to have any impact it must be followed up by further reports or better yet by political or legal action.

Some scientists have succeeded in becoming recognized sources of news by banding together to form organizations like the Colorado Committee for Environmental Information and establishing a reputation for accuracy and newsworthiness, or else by working through established scientist "front" organizations, like the Federation of American Scientists, that already have such a reputation. An alternative is to seek support from recognized citizens' groups like the Sierra Club or Friends of the Earth, or perhaps to seek the assistance of Ralph Nader—as the Union of Concerned Scientists have done in their campaign for increased reactor safety. The traditional device of the petition, which was used by Meselson and his colleagues in calling for a Presidential reexamination of U.S. chemical and biological warfare policies, has fallen somewhat into disuse. It is now associated with quieter days, when policy for technology was relatively uncontroversial and it was a newsworthy event when a dozen Nobel Prize winners or a few hundred ordinary scientists disagreed enough with established policy to sign their names on a sheet of paper. Since it has almost become the norm for the majority of the population to disagree with established policy, more substantial protests are required to gain serious public attention.

LEADING THE WAY

In between the "popular press" and the scientific journals lies a third category of magazines, edited by scientists but aimed at scientists and laymen alike. Most notable among them are *Science, Scientific American,* the *Bulletin of* the *Atomic Scientists,* and *Environment* in the United States and *Nature* and *New Scientist* in Great Britain. Some of the articles in these magazines (in the case of *Environment,* all the articles) relate to policy issues concerning technology. *Science,* in addition to publishing occasional articles from outside contributors on such subjects, has a full-time staff which concentrates on reporting on current controversies in the science and technology area. Articles in these magazines have played a crucial role in making debates on many technological issues accessible to the popular press. Often such an article has served to establish the credibility and importance of dissenting views on a particular issue, inasmuch as it is recognized that the article will have been reviewed by competent scientists, including the editors, who would presumably have rejected it if it were obviously in error or overly speculative.

The 1968 *Scientific American* article by Bethe and Garwin on the Sentinel ABM system is a notable example.[4] It explained, using nonclassified information but nevertheless in a specific way, how the Johnson administration's proposed antiballistic missile system could be penetrated by enemy missiles with relative ease. This article had a substantial effect in convincing other, previously uninvolved members of the scientific community that the ABM system, besides further escalating the arms race, would be a terrible waste of money and would become more and more expensive as the Defense Department tried to compensate for its intrinsic weaknesses. Many of these newly persuaded scientists then carried the issue to the public and to Congress.

The articles in the "News and Comments" section of *Science* have become steadily more important in bringing serious problems to public attention. For example, a series of investigative articles by *Science* reporter Robert Gillette on the nuclear reactor safety issue[5] effectively made that subject accessible to the press and probably played a crucial role in the later firing of AEC nuclear reactor czar Milton Shaw and the restructuring of his former empire. In another case, scientists muttered about "blacklisting" by the Department of Health, Education, and Welfare for years, and twenty professional societies even joined to petition HEW privately to discontinue the practice, but nothing happened until Bryce Nelson made the issue public in a series of articles in *Science*.[6] By obtaining a list of forty-eight blacklisted scientists, including one Nobel Prize winner, Nelson established that the blacklisting was actually a reality. In the six months following Nelson's first article in June 1969, more than a hundred articles and critical editorials appeared in newspapers and periodicals across the nation, and the issue was even discussed on network television. Congressional pressure developed—Senator Sam Ervin (D.-N.C.) twice emphasized to HEW Secretary Robert Finch that blacklisting is a "violation of constitutional principles which cannot be tolerated"—and in January 1970 HEW decided to abandon the practice.[7]

TALKING TO REPORTERS

There is little admiration lost between most reporters and most scientists. To reporters, scientists often seem preoccupied by details, and unable to communicate what is really bothering them. On the other hand, scientists too frequently find that reporters miss the real point and can be restrained only by force from rushing off to publish a completely misleading story. Obviously, both sides must work to close the gap.

One might add the observation that papers with well-educated readerships like the *New York Times* and *Los Angeles Times* have sophisticated reporters who are ordinarily given more time to work up a story than the reporter on your local *Daily Advertiser*. In this connection the Colorado Committee for Environmental Information initially found it easier to get coverage in the national media than in the local Colorado papers. Peter Metzger summed it up with the biblical observation: "A prophet is without honor in his own land."[8] Finally, when dealing with the ordinary reporter, who has probably just returned from filing a story on a former poetry teacher who took off her bikini top in the center of the financial district, there is obviously no substitute for a brief, well-written press release containing the essential information.

NOTES

1. Tom Wicker, "The Reporter and His Story: How Far Should He Go?" *Nieman Reports,* Spring 1972, p. 15.

2. *New York Times,* March 1, 1971, p. 15.

3. Peter Sandman has pointed out that the press is much more effective in telling us what to think about than it is in telling us what to think. See "Mass Environmental Education: Can the Media Do the Job?", to be published in *Environmental Education,* William B. Stapp and James A. Swan, eds. (Beverly Hills, Calif.: Sage Publishing Company, 1974).

4. Hans Bethe and Richard Garwin, "Anti-Ballistic Missile Systems," *Scientific American,* March 1968, pp. 21-31. See Chapter 13.

5. See note 43, Chapter 15.

6. Bryce Nelson, "HEW: Finch Tries to Gain Control Over Department's Advisory Groups," *Science* 164 (1969): 813. See also *Science* 165 (1969): 269, and *Science* 166 (1969): 357, 819, 1488.

7. James Singer, "Pressure by Scientists, News Media Figured in HEW Curb on Blacklisting," *National Journal,* January 10, 1970, p. 73. See also Bryce Nelson, "HEW: Blacklists Scrapped in New Security Procedures," *Science* 167 (1970): 154.

8. Peter Metzger, "The Colorado Committee for Environmental Information," *Report of the Conference on Scientists in the Public Interest: the Role of Professional Societies* (Salt Lake City: Physics Department, University of Utah, 1974). (The conference was held at Alta, Utah, September 7-9, 1973, under the sponsorship of the American Academy of Arts and Sciences, Western Center, and the University of Utah.)

CHAPTER 17

Organizing for
Public Interest Science

Traditionally, public interest science has been an activity carried on in an entirely ad hoc manner by full-time scientists and engineers who have taken time off from their usual pursuits. They don their white hats and gallop off to rescue imperiled Paulines just as doom seems imminent—and then they return to the laboratory.

It is important that such "amateur" public interest science continue. Until recently the scientific community delegated its public responsibilities mostly to official government science advisors. This was a mistake. As the histories of government regulatory agencies have repeatedly demonstrated, responsibility cannot successfully be delegated—it can only be shared. The unfettered spirit of part-time outsiders will always be required to keep the system honest.

But neither is a system in which public interest science is practiced only by volunteers satisfactory. Nothing less than a full-blown crisis is required to motivate a dedicated scientist to drop his usual work. By that time, it may be rather late to initiate corrective steps. It would have been far better, for example, if the adequacy of the AEC's reactor safety program had been subjected to independent review a few years earlier. This would have saved the large amounts of money which may be required to fit existing reactors with improved safety systems and would have reduced the risk—whatever it may be—to those persons who will be living near those reactors in the meantime.

In most of our examples of independent public interest science activities—regarding DDT, plutonium and nerve gas in Colorado, defoliation in Vietnam, and so on—independent scientists reacted only after years of government misconduct of technological programs. It should by now be obvious that if the public interest is to be adequately represented in governmental decisions on technological issues, public interest science must to some degree be institutionalized.

Institutionalizing the outsider role poses a great challenge to the creativity of scientists and the scientific community. Funding is obviously required for such an effort, but the customary sources of funding for scientists—the federal government and industry—are just those institutions whose policies may have to be challenged. Even non-mission-oriented government agencies like the National Science Foundation (NSF) have been very reluctant to support "controversial" public interest science projects or groups, although controversy is often essential to bring out all the important considerations in governmental decisions. Thus in 1971 NSF refused to support the activist-oriented but responsible magazine *Environment,* while at the same time continuing to fund the noncontroversial (and rather dull) *Science News.* [1]

Fortunately, private foundations are beginning to show some interest in funding public interest science. The Ford Foundation, for example, sponsored the wide-ranging Energy Policy Project in 1972-1973 and has for several years provided partial support to groups like the Environmental Defense Fund. The Stern Fund contributes to the support of the new Center for Science in the Public Interest in Washington, D.C. Federal and state governments may yet decide to fund public interest science projects as the field becomes more respectable—like public interest law.

The more fundamental problems of public interest science are thus likely to lie less in the area of funding than in the professional motivations of and institutional constraints on scientists. In this chapter we will first consider the nature of these constraints and then examine some of the ways in which scientific professional societies and public interest organizations can organize— and to a certain extent are already organizing—public interest science activities.

Scientists

INDUSTRIAL SCIENTISTS

Corporate employees are among the first to know about industrial dumping of mercury or fluoride sludge into waterways, defectively designed automobiles, or undisclosed adverse effects of prescription drugs and pesticides. They are the first to grasp the technical capabilities to prevent existing product or pollution hazards. But they are very often the last to speak out. [2]

—Ralph Nader

Most scientists and engineers are employed in industry. There they are perfectly situated to see first-hand the potential and real hazards of industrial products and practices and to suggest steps to remedy them. But few industrial scientists speak out, even within the corporate hierarchy. Advancement comes to those whose work pays off in increased corporate profits (and sometimes to those who just put in their time); career stagnation or termination is the usual reward for "troublemakers." The First Amendment protects the right of free speech only from governmental interference; private employers are not bound by it. Unless constrained by law—as in the federal antidiscrimination statutes—or by an explicit employment contract, any company can deal with its employees in an essentially arbitrary manner. Although industrial unions have won a variety of rights for blue-collar workers, few industrial scientists or engineers have even the most elementary employment safeguards. Indeed, their contracts, if they have any, are often replete with provisions intended to discourage independent action. Such provisions can apply even after retirement: Du Pont warns its retirees that their pensions can be canceled if they engage in "any activity harmful to the interest of the company."[3]

GOVERNMENT SCIENTISTS

Government employees would at first sight appear to be much better protected than corporate employees, since they benefit from both Constitutional and Civil Service safeguards. But the harassment and eventual firing in 1969 of A. Ernest Fitzgerald, the Pentagon cost analyst who revealed the cost overruns in the manufacture of the Air Force's C-5A transport, shows how limited these protections can be. (Fitzgerald was ultimately reinstated by the Civil Service Commission with three years' back pay because some memos surfaced during the Watergate investigation which allowed him to prove what everyone knew—that considerations entered into the abolition of his job other than those of "economy.")[4] Like other large bureaucracies, government agencies reward quiet mediocrity more regularly than aggressive pursuit of the public interest. That may be why none of the industrial or government scientists who were aware of the Bionetics Research Laboratories findings on the teratogenicity of 2,4,5-T spoke up during the three-year period while the information was being suppressed. And why the reactor safety issue festered quietly within the AEC for so many years before it was brought out into public view.

Efforts can be made to intimidate a government employee even when he is not criticizing his own agency. For example, during the controversy over the plutonium pollution outside Dow Chemical's Rocky Flats plant in Denver, Dr. Martin Biles, director of the AEC's Division of Operational Safety, approached Robert Williams and Dion Shea of the Colorado Committee for Environmental Information and informed them that he had a "personal hangup about one federal agency engaging in activities critical of another federal agency," adding: "You don't mind if I bring this matter up with the appropriate officials of [the Department of] Commerce [their employer] and the National Science Foundation [which funded the research of Edward Martell, the scientist who had done the CCEI plutonium measurements]."[5]

UNIVERSITY SCIENTISTS

As the world becomes more technically unified, life in an ivory tower becomes increasingly impossible. Not only so; the man who stands out against the powerful organizations which control most of human activity is apt to find himself no longer in the ivory tower, with a wide outlook over a sunny landscape, but in the dark and subterranean dungeon upon which the ivory tower was erected. . . . It will not be necessary to inhabit the dungeon if there are many who are willing to risk it, for everybody knows that the modern world depends upon scientists and, if they are insistent, they must be listened to.[6]

—Bertrand Russell

Thus we come to the universities. University scientists, protected by a long tradition of academic freedom, are in principle free to speak their minds and take public stands on any issue. And indeed many of the scientists whose public interest activities we have discussed have been affiliated with universities.

The majority of university scientists, however, have remained entirely uninvolved in public debates about technological issues. And of those who have forsaken the ivory tower for such activities, the number consulting for government or industry has been far larger than the number of independent public interest scientists.

One reason for this lack of involvement in public interest activities appears to be the fact that after World War II the university changed from a haven for poorly paid and rather solitary teachers and researchers into a busy confluence of traffic in the high-pressure world of advanced technology. The established academic scientist now typically administers a research group supported by several annually renewed government or industrial research contracts and is continually concerned that his group's output be of sufficiently high caliber to insure that its funding will be renewed or (hopefully) expanded. He makes frequent trips to Washington in search of funds and in his capacity as a government advisor. He attends conferences all over the world where he tries to make sure that the accomplishments of his group are visible and acknowledged. Finally, he usually also teaches a course at the university and supervises the work of several graduate students.[7] Rising younger scientists lead a somewhat less frenetic existence, but they are generally working overtime on scientific problems that interest them, establishing their own professional reputations, and competing to emulate their senior colleagues.

With such demanding professional lives, it is not difficult to understand why academic scientists have not been very open to the challenges of public interest science. Not only would such activities distract the scientist from his efforts to preserve and enhance his own and his group's position in the highly competitive world of scientific research, but they also might result in his being labeled a "controversial figure," an image that could adversely affect the delicately balanced judgments on which promotion and funding decisions are often based. None of these problems is likely to afflict a scientist who minds his own business or only consults privately for industry and the government.

Fortunately, in recent years the rigidity of these traditional professional patterns has shown signs of weakening as the scientific community has begun to recognize that the era of almost unquestioning faith in science and technology, which began with the development of the atomic bomb and was sustained for a time by the challenge of *Sputnik,* has come to an end. The nation's primary concerns have finally turned from security against external threats to enhancing the quality of life at home. And here the public has discovered that many of the new devices and chemicals that technology has been constantly producing for the domestic market have a serious potential for damage to human and environmental health. Technological time bombs have begun to explode: smog, destruction of entire wildlife populations by DDT, jet noise near metropolitan airports, and the suspicion that birth defects and cancer may be linked to the new substances to which man has exposed himself in his work, environment, and food. A "backlash" against technology has developed. And many scientists have become genuinely concerned about ameliorating the adverse consequences of technology and regaining the respect of the public—including their students, families, and friends. The strong constraints imposed by professional ambition still exist, but attitudes within the technical community are changing from skepticism of public interest science activities toward neutrality and perhaps even a certain amount of encouragement. These changes are manifested in the new social activism of many scientific professional societies, the recent birth or rein-vigoration of several public interest science groups, and the steadily increasing number of full-time public interest scientists.

Professional Societies

Traditionally, scientific professional societies have restricted their activities to the sponsorship of professional meetings and the publication of technical journals—i.e., to the discussion of developments in their respective areas of specialization—sometimes also awarding honors to members who have made notable scientific advances. This single-mindedness has been defended as a virtue by the leaders of various societies, who are concerned lest discussion of "political" matters such as the social impact of the applications of their field polarize their membership, pollute their discipline, and generally bring scientists and the supposed objectivity of the scientific method into disrepute. The common attitude has been that scientific discussion should be strictly segregated from the discussion of questions which cannot be answered using the scientific method and that the scientific societies, as the inner sanctums of the scientific enterprise, need special protection.

As concern has increased in the country over the adverse impacts of technology, however, scientific societies have found it more and more difficult to remain uninvolved. Recent unemployment problems have also led scientists to demand that their professional societies undertake a number of new activities—ranging from employment information services to outright lobbying for more federal support for science. Both because of the job crisis and because of general dismay over technological fiascos like the ABM and SST, the interaction of science with society has come to be recognized as a legitimate concern of scientists as professionals. Although the defenders of the traditional aloofness of professional societies have urged those who feel compelled to discuss the ways that science impinges on society to find another forum, the inescapable fact is that there neither are nor ever have been other comparable forums for such discussions within the scientific professions.

PROFESSIONAL SOCIETIES AS SPONSORS FOR PUBLIC INTEREST SCIENCE

For the coming decade the main thrust of AAAS attention and resources shall be dedicated to a major increase in the scale and effectiveness of its work on the chief contemporary problems concerning the mutual relations of science, technology, and social change, including the uses of science and technology in the promotion of human welfare.[8]

—Board of Directors,
American Association for the
Advancement of Science, 1969

The leader among professional organizations in science and society issues has been the 130,000-member American Association for the Advancement of Science (AAAS), the publisher of *Science* magazine. Although it is an organization dominated by scientists, AAAS is not itself a professional society, but rather a loose association of virtually all of the 300 specialized scientific and engineering societies in the United States. Since the early 1950s, the AAAS has increasingly concerned itself with public issues, leaving the work of furthering the development of each discipline to the more specialized societies.

In the constitution adopted by the AAAS in 1946, one of the principal goals of the organization was stated to be improving "the effectiveness of science in the promotion of human welfare." But the AAAS moved to implement this goal without noticeable haste. A decade and a half later, a Committee on Science in the Promotion of Human Welfare was appointed to look into the matter.

The committee decided that the single most important way in which scientists can help society solve the problems that have been created by scientific advances is by informing their fellow citizens of the relevant facts. "In sum," stated their first report,

we conclude that the scientific community should on its own initiative assume an obligation to call to public attention those issues of public policy which relate to science, and to provide for the general public those facts and estimates of the effects of alternative policies which the citizen must have if he is to participate intelligently in the solution of these problems. A citizenry thus informed is, we believe, the chief assurance that science will be devoted to the promotion of human welfare.[9]

STUDIES OF PUBLIC ISSUES

Thus far, the most venturesome study sponsored by the AAAS—or, for that matter, by any scientific professional organization—has been that of the Herbicide Assessment Commission. As we recounted in Chapter 11, the initial impetus for this project came in 1966 from E. W. Pfeiffer, a Montana zoologist who was also one of the founders of the Scientists' Institute for Public Information. The AAAS leadership timidly resisted involving their organization in this highly charged issue for more than three years—years during which the Army conducted the bulk of its defoliation operations. But when the project was finally undertaken under the leadership of Matthew Meselson, the work of the Herbicide Assessment Commission was of such unimpeachable quality and its conclusions so carefully stated that it has reflected nothing but credit on the AAAS. And the undertaking had great political impact—the photographs with which the HAC returned from Vietnam brought home to the American people the devastation being caused by the defoliation program and helped to bring about its termination.

The 110,000-member American Chemical Society (ACS) has been the pioneer among specialized professional societies in preparing public reports on technical issues. In 1965, inspired by the President's Science Advisory Committee report *Restoring the Quality of Our Environment,*[10] the ACS Committee on Chemistry and Public Affairs, in cooperation with the ACS Division of Water, Air, and Waste Chemistry, recruited a panel of experts to prepare a handbook on pollution that would be suitable for Congressmen and other interested laymen. They received encouragement from President Johnson's science advisor, chemist Donald Hornig, and a number of Congressional leaders—but they were also cautioned by these men to avoid bias in favor of the chemical industry.

The experts were assembled for a two-day meeting in the expectation that the report could be drafted in one or two such sessions. What resulted, however, according to Stephen Quigley, ACS Director of Chemistry and Public Affairs, was "a veritable Tower of Babel."[11] Finally, after much more work than initially anticipated, a first draft of the report was finished. But it was intelligible only to scientists, so it was sent back for redrafting. Eight revisions later the steering committee agreed that it was both suitable for general consumption and scientifically sound. The report, *Cleaning Our Environment: The Chemical Basis for Action,* was finished in 1969.

All this work did not go unrewarded. More than 50,000 copies of *Cleaning Our Environment* have been sold to the general public and to students—in addition to the 21,000 copies that were initially distributed to federal, state, and local officials and to the news media. The report has been used in some 130 colleges as a textbook.

PROFESSIONAL SOCIETIES AS FORUMS FOR PUBLIC ISSUES

Most professional societies have been much less active in studying public issues than the AAAS or the ACS. Of those that have been involved at all, the majority have confined their activities to sponsoring talks and panel discussions at their meetings. That has been the main function, for example, of the Forum on Physics and Society of the 30,000-member American Physical Society (APS). At all major APS meetings during the past several years, the Forum on Physics and Society has sponsored programs on a very wide range of subjects, including the antiballistic missile debate, pollution problems, population and economic growth, problems of women and other minorities in physics, secrecy in science, Soviet scientists and human rights, and the employment crisis in physics. These sessions have almost always been very well attended.

It is very important that the opportunity exist for discussion of such issues among scientists. Ordinarily, when a new technological program is being "sold" to the executive branch (e.g., the ABM or other weapons systems, the SST, the "breeder" nuclear reactor, the "war" on cancer, etc.), discussion is pretty well confined to that part of the technical community most closely tied to the industries and/or government agencies involved. This results in troublesome issues "sleeping" long after they should have been brought out into the open and resolved. For example, if the very great psychological impact and substantial physical destructiveness of sonic booms from the heavy U.S. SST had been as widely understood in 1964 as they were in 1969, a much sounder basis for discussing and planning the SST program would have existed. There was no good reason why the seriousness of these problems and their intractability to any kind of "technical fix" could not have been made clear several years earlier than they were. Another such example is nuclear reactor safety: if the adequacy of the safety systems had been critically reviewed by the larger technical community before construction on the present generation of large power reactors was begun, the AEC and the electric utility industry might have been spared a lot of grief.

Through such institutions as the APS Forum on Physics and Society, it should now be possible for concerned individuals—such as Shurcliff in the case of the SST or Kendall on reactor safety—to raise important issues regarding the effects of the proposed technology in front of a disinterested but nevertheless competent group of scientists. Ideally, such discussions should take place long before issues reach the crisis stage. In cases where there is substantial disagreement over either the facts or their implications, more sustained and serious inquiry should be possible. For example, professional societies, either individually or jointly, could sponsor meetings or topical conferences at which all interested scientists would be able to discuss their views and clarify specific areas of disagreement. Or, in complex areas such as reactor safety, prolonged studies might be organized—over the summer, presumably, in obeisance to the academic calendar. The results of such efforts would surely be useful to both the executive and legislative branches of the federal and state governments and to all citizens who are concerned about these issues.

Traditionally, the scientific community has assumed that if such studies were needed, they would be undertaken by an executive-branch science advisory committee or by the National Academy of Sciences' National Research Council. It is important to appreciate, however, that these bodies are "other-directed," not "inner-directed"—i.e., they usually respond to requests from executive agencies rather than initiating studies on their own.[12] And as we have seen, such studies are vulnerable to suppression or subversion by the sponsoring agency.

Primary responsibility, therefore, remains with the larger scientific community to help identify and call public attention to the crucial questions and to see to it that necessary studies are performed. If the government is willing to arrange for open, high-quality studies—fine. But if not, the professional societies should be prepared to organize them on their own, as the AAAS finally did when it established Herbicide Assessment Commission.

FACILITATING PUBLIC INTEREST SCIENCE

We have been arguing that in addition to its usual function of advancing and diffusing the knowledge of its particular discipline, a professional society can also provide a unique scientific forum for the discussion and study of public issues with technical components. Indeed, professional societies represent among their members the collective scientific wisdom and knowledge of the nation. The higher officials of the federal executive branch can call upon this expertise through science advisory committees and the National Academy of Sciences. State and local governments could in principle go this route—and some have tried—but they usually lack the dual concentrations of responsibility and expertise which have made such arrangements successful at the federal level. Most citizen and public interest groups have the additional problem that they do not have the resources for formalized consulting arrangements.

So where does a governor turn when he wants independent advice about the potential safety problems of a new nuclear reactor or tank farm for liquified natural gas under construction in his state? Or if a committee of the state legislature wants to know how privacy of information can be protected in the state's computerized data banks, whom does it consult? (The local chapter of the American Civil Liberties Union may have the same question.) Or, again, where does the St. Louis People's Coalition Against Lead Poisoning go if it wants to know how to determine whether the paint peeling off a particular wall has a lead-based pigment? Access to names of executive-branch advisors will not be enough—if only because the group seeking advice may be in an adversary relationship with the federal agency or because many of the well-known scientists who fly off to Washington to consult would not have enough time in their busy professional lives to advise the mayor, state assemblyman, and local chapter of the Sierra Club as well.[13]

In many instances, however, scientists with the necessary competence would be delighted to help. The problem is to get the willing scientist together with the interested official or citizens' group. Scientific societies can help fill this need by making easier the connection between groups which need advice and qualified scientists interested in participating in public interest advisory activities.

The Biophysical Society is pioneering in setting up a system for such "matchmaking." The scientists in this small (2,500-member) society possess expertise which is especially relevant to determining the subtle biological effects of radiation, food additives, and chemical pollutants. Just before assuming the society's presidency in 1972, Peter von Hippel (brother of one of the present authors) sent out questionnaires to the membership asking whether the Biophysical Society should

participate in an organized form in making available and providing scientific advice to the various branches of federal, state, and local government and to citizen groups.[14]

The proposal was approved by a ten-to-one margin. Accordingly, a committee was appointed to prepare a detailed computer-compatible questionnaire by which members could indicate the technical areas in which they were willing and competent to do public interest work.

Much thought went into how the program should operate. The model finally chosen was that of an "editorial board" of experts who would receive requests for assistance in their various areas of expertise and would then be responsible for selecting advisors from the Biophysical Society's roster and initiating contact between the advisors and their "client." The "editor" assigned to a particular request would receive copies of any reports prepared and might append his own comments if he felt this to be helpful or appropriate. It is anticipated that any costs for travel, secretarial help, and the like would be borne by the "client" individual or group; in exceptional cases the society might try to find an alternative source of funds or provide partial support from its own funds.

There are several reasons for interposing an editorial board between advisors and their prospective clients. Besides helping to find the best advisor for each request and monitoring the subsequent advisory relationship, the editorial board would also serve to screen out inappropriate requests. (For example, Peter von Hippel tells of one request from a lawyer in Wisconsin whose client had hurt herself in a fall and was suing for damages. The lawyer's request?—a complete list of all possible injuries his client might have suffered!) The following statement was decided upon to help determine the appropriateness of requests:

The basic purpose of the advisory service of the Biophysical Society is to contribute to the improvement of conditions of society . . . to relieve suffering and prolong life, to improve the environment by reducing pollution of the air or water or protecting natural resources . . . .[15]

It was also decided that the editorial board would retain, and generally exercise, the option of making the results of investigations public.

Peter von Hippel reported on the progress of the Biophysical Society's public advisory project at a conference, Scientists in the Public Interest: The Role of Professional Societies, held in Alta, Utah, in the fall of 1973.[16] The enthusiastic response of the other participants, including representatives of a number of professional societies, indicates that other societies may soon join the biophysicists in offering their services to the public. Such services could also prove helpful to officials responsible for choosing members for science advisory committees organized by federal agencies or the National Academy of Sciences.

DEFENDING PROFESSIONAL RESPONSIBILITY

Until recently, all hopes for change in corporate and government behavior have been focused on external pressures on the organization, such as regulation, competition, litigation, and exposure to public opinion. There was little attention given to the simple truth that the adequacy of these external stimuli is very significantly dependent on the internal freedom of those within the organization.

... Within the structure of the organization there has taken place an erosion of both human values and the broader value of human beings as the possibility of dissent within the hierarchy has become so restricted that common candor requires uncommon courage.

There is a great need to develop an ethic of whistle blowing which can be practically applied in many contexts, especially within corporate and governmental bureaucracies. For this to occur, people must be permitted to cultivate their own form of allegiance to their fellow citizens and exercise it without having their professional careers or employment opportunities destroyed. ... Whistle blowing, if carefully defined and protected by law, can become another of those adaptive, self-implementing mechanisms which mark the relative difference between a free society that relies on free institutions and a closed society that depends on authoritarian institutions.[17]

—Ralph Nader

Another topic much discussed at the Alta Conference on public interest science was the role of professional societies in defending the professional integrity of scientists. A number of professional societies have included relevant passages in their professional codes of ethics. Thus we find in the code of the National Society of Professional Engineers:

The Engineer will have proper regard for the safety, health, and welfare of the public in the performance of his professional duties. If his engineering judgment is overruled by non-technical authority, he will clearly point out the consequences. He will notify the proper authority of any observed conditions which endanger public safety and health.[18]

And the *Chemist's Creed* of the American Chemical Society contains the following:

As a chemist, I have a responsibility ... to discourage enterprises or practices inimical to the public interest or welfare, and to share with other citizens a responsibility for the right and beneficent use of scientific discoveries.[19]

But most scientists and engineers have heavy family responsibilities and are locked into their jobs by the uncertainty of whether they could find another comparable position without an intervening period of severe dislocation. To them, therefore, the high-sounding phrases in their professional codes of ethics must seem pretty remote.[20] If scientists and engineers felt that their professional societies would stand behind them when they acted according to these codes of ethics, things might be somewhat different.

There are many things that scientific societies can do to defend the professional integrity that their codes of ethics urge upon their members. At the very least they can lobby for legal protection for the government or industrial professional who refuses to carry out orders which violate either the letter or spirit of the law or imperil the public health and safety. Professionals should have legal protection against losing their means of livelihood as a result of actions in the public interest, or at least they should be able to sue for compensation and expect a timely hearing of their suit.

Until our legal system recognizes the value to the public interest in offering protection to "whistle blowers," professional societies must fill the gap to the extent that they are able. The American Association of University Professors (AAUP) works to protect the academic freedom of its members by setting certain standards for the universities at which they are employed. When it appears that a university's treatment of one or a number of its faculty members has violated these standards, the AAUP often conducts an inquiry on the basis of which, in extreme circumstances, it may publicly censure the university. There is no reason why professional societies cannot involve themselves in similar activities in defense of the professional integrity of their members. In those cases where a society fails to dissuade an employer from seeking revenge on a whistle blower, the society could exert itself to help him find new employment and even provide legal assistance in a suit against his former employer if both the society and the member feel that the case has sufficient merit. Very few such cases have ever been taken to court, but a few well-chosen litigations could establish landmark precedents.

President Alan C. Nixon of the American Chemical Society reported at the Alta Conference that the ACS has undertaken essentially all of the activities mentioned above. It has established a professional relations committee to develop model employment contracts and investigate members' employment grievances and a legal aid fund to act on the professional relations committee's findings if necessary. The ACS also plans to compile an annual publication listing the employment practices of the 900 leading employers of chemists, including records of member complaints and ACS findings.

It was also suggested at the Alta Conference that the societies recognize notable accomplishments in public interest science just as they hand out awards for notable scientific discoveries:

> In order to strengthen the general respect for professional codes of ethics, societies could . . . give certificates of commendation to individual scientists whose integrity has defended the public health and welfare against significant hazards as in the famous case of the FDA medical scientist, Dr. Frances Kelsey, who held the line on Thalidomide.[21]

In this vein, the APS Forum on Physics and Society in 1974 established the Leo Szilard Award for Public Interest Science. The first recipient was David R. Inglis.

Public Interest Science As a Profession

Until the late 1960s, debates over technology generally focused on particular dangers of particular technologies: the side effects of drugs such as thalidomide, the dangers of fallout from atmospheric nuclear testing, the dangers of persistent pesticides, and so forth. In the past few years, however, the public has come to recognize that almost all technologies have potentially adverse side effects. The response has been to try to develop institutions and laws which set up mechanisms for the determination and regulation of the impact of technologies in general rather than continuing to react to problems on an individual and ad hoc basis. Thus we have the National Environmental Protection Act (1969) with its requirements of "environmental impact statements" for federally funded or regulated projects, the Environmental Protection Agency (1970), and Congress's new Office of Technology Assessment (1973).

The public interest science movement is also starting to institutionalize. As yet, the number of professional—i.e., full-time—public interest scientists is very small. We will discuss a few of these pioneers briefly here.

RALPH LAPP

Dr. Ralph Lapp is a "free-lance" public interest scientist: he works alone and with no organizational base. Lapp worked on nuclear weapons during the Second World War and on the development of nuclear reactors for a few years thereafter. Since about 1950, however, he has been an independent and respected critic of U.S. policy in these areas. Lapp's first great success in his new career was with his book *The Voyage of the Lucky Dragon,* the true story of an unlucky Japanese fishing vessel which was caught in the radioactive cloud from one of the United States H-bomb tests in the South Pacific.[22] This best-selling book helped bring home to the public the hazards of fallout from atmospheric nuclear weapons tests. More recently, Lapp has participated in the debates over the deployment of antiballistic missiles and the safety of nuclear reactors. He has written many books on issues relating to the arms race and most recently on the "energy crisis."[23] Many of his articles have appeared in the *New York Times Magazine* and in the *New Republic.*

Lapp supports himself by his writing, by giving talks to university and industrial groups, and as a consultant (in 1972, for example, to state officials concerned about the safety of nuclear reactors being sited in their jurisdictions). He prefers to act as a friendly critic of the AEC. As a result, he has good communications with the AEC's Commissioners and high-level bureaucrats, and he tries to influence policy through this access route both before and during public debates over AEC policies. He has been quite effective at this—perhaps because he has demonstrated that he is willing and able to take issues to the public when he thinks it is necessary.

JEREMY STONE

Still in his thirties, Dr. Jeremy J. Stone abandoned a promising academic career in 1970 to become the first full-time executive director of the Washington, D.C.-based Federation of American Scientists (FAS) since that organization's beginnings in 1945-1946. (The FAS had been born in the post-war scientists' campaign for the assignment of responsibility for atomic energy to an agency under civilian control.[24]) In the late 1960s at about the same time a substantial number of high-level government science advisors began to move outside government and work through the FAS—partly in order to bring before Congress the ABM debate which they had lost within the executive branch in 1967, also as a result of their frustration with the Indochina war, and finally—perhaps most importantly—because of their diminishing influence within the Johnson and Nixon administrations. The FAS welcomed the support of these former insiders, and by 1972 a former head of the elite Jason group of Defense Department consultants, Marvin Goldberger, had succeeded former Director of Defense research and Engineering Herbert York in the (unsalaried) FAS chairmanship. Partly as a result of the support of these prominent figures, and partly because of Jeremy Stone's dedicated and imaginative leadership, the FAS has experienced a considerable reinvigoration.

Stone's efforts were crucial in convincing the Armed Services Committees of both Houses of Congress to institute a new tradition of inviting witnesses opposed to administration proposals to hearings on weapons systems. And the testimony which he has organized against the Pentagon's favorite new weapons boondoggles, an effort that has sometimes pitted former high executive-branch officials against the current occupants of the same offices, has not been without effect. For example, FAS witnesses helped convince Chairman John Stennis (D.-Miss.) of the Senate Armed Services Committee to refuse flatly the Nixon administration's 1971 request to expand the Safeguard ABM system. In recent years FAS has developed positions on a broad spectrum of technological issues—the SST, reactor safety, world food supply, ways of reducing air pollution from automobile emissions, the oil crisis, and so forth—and its monthly newsletter, renamed in 1973 the *FAS Public Interest Report,* has become a steadily improving digest of informed scientific opinion on controversial issues. (In writing and editing this newsletter, Jeremy Stone adheres to the sort of independent journalism his father pioneered in *I. F. Stone's Weekly.*) As a consequence of its new record of accomplishment, coupled with Jeremy Stone's indefatigable campaigns to attract new members, the FAS's membership tripled over a recent two-year period and reached a total of about 6,000 in 1973.

JAMES MACKENZIE

Dr. James MacKenzie, a nuclear physicist, also in his early thirties, became involved in public interest science as one of the leaders of the Union of Concerned Scientists and director of UCS environmental activities. In 1970-1971 the group lobbied the Massachusetts Department of Public Health—first in favor of setting air quality standards for the Boston area and then in opposition to Boston Edison's request for a variance from these standards for a large coal-burning power plant.

During this effort the UCS became disgusted with the proindustry bias of the Public Health Department and its relative insensitivity to threats to the public health. As a consequence, MacKenzie and his group prepared and distributed a Ralph Nader-type exposé on the pesticide-regulation, air-pollution control, and meat-inspection policies of the department that ultimately led to the governor's replacing several top state health officials with men more interested in public health.

In 1970 MacKenzie took a full-time position with the Massachusetts Audubon Society, where he has since established himself as an "environmental scientist" and as a nationally recognized generalist on energy technology. He is much sought after to serve on federal executive-branch advisory panels, he has become increasingly active as an advisor to Massachusetts state officials, and he continues his public interest work. In 1972, together with James Fay, a professor of mechanical engineering at MIT, MacKenzie called increased public attention to the dangers associated with the unloading and storage of liquified natural gas near metropolitan areas. They explained that if a large tanker or storage tank should rupture, it would release a large cloud of cold vapor which would drift along the ground ready to ignite. The resulting fire could incinerate more than a square mile.[25] MacKenzie also has a special interest in solar energy and has persuaded the Massachusetts Audubon Society to advance the state of the art by designing its new office building to be both heated and cooled using this energy source.

THE EDF SCIENTIFIC STAFF

In 1971 the Environmental Defense Fund began hiring young scientists to complement the increasing number of lawyers on its professional staff. Leo Eisel, a water-resource and land-use-planning engineer who had worked as a student with Jim MacKenzie and the Union of Concerned Scientists, was one of the first of the full-time EDF scientists. By 1973 the scientific staff of the EDF had grown to six and the scientific preparation of many of the organization's cases was being handled primarily by these scientists. Meanwhile, in other areas of activity—particularly pesticides—the traditional part-time public interest scientists such as Charles Wurster continued to pull their weight.

THE CENTER FOR SCIENCE IN THE PUBLIC INTEREST

Our final example of the professionalization of public interest science is the Washington, D.C.-based Center for Science in the Public Interest.

Dr. Albert Fritsch (an organic chemist and Catholic priest), Dr. Michael Jacobson (a biochemist), and Dr. Jim Sullivan (who is trained in meteorology and oceanography) all began their public interest careers by working for Ralph Nader. In January 1971 they incorporated as the nonprofit, tax-exempt Center for Science in the Public Interest (CSPI) for the purposes of:

1. Collecting and publicizing evidence to assess whether public and private activities involving technology are truly reponsive to the public interest;

2. Encouraging scientists and engineers working in government and industry to be more aware of citizen needs; and

3. ... Promoting legal action or administrative appeals, supplying legislatures with requested data, or focusing public pressure on critical and consumer issues.[26]

Mike Jacobson has specialized in food additives. His popular writing on the subject has been quite well received: his book *Eater's Digest*,[27] written while he was still with Nader, had sold more than 25,000 copies by the summer of 1973, and his pamphlet, *Nutrition Scoreboard*, was then selling at the rate of 250 orders a day. In addition, Jacobson has written a number of more specialized reports on particular problems, including a pamphlet on sodium nitrite (entitled *Don't Bring Home the Bacon*) and one on *The Chemical Additives in Booze.*[27] As a result of Jacobson's activities in connection with the latter topic, the Internal Revenue Service in 1973 issued a ruling that the chemical additives in beer, wine, and hard liquor must be listed on the labels, as they are for food.

Two of Al Fritsch's projects have involved gasoline additives and asbestos pollution, and he has written several reports on these subjects. In the case of asbestos fibers, which are known to cause lung cancer, Fritsch has been pressing all the responsible federal agencies to act in their areas of responsibility in the expectation that their actions will be mutually encouraging and reinforcing. Regarding gasoline additives, his concern is that some of the additives may give rise to dangerous (e.g., cancer-producing) air pollutants. He has managed to persuade the Environmental Protection Agency to release a list of two-thirds of the additives in gasoline and has initiated a suit to obtain the rest. In response to the claim that this information involves trade secrets, the CSPI contended that the oil companies could always chemically analyze each other's products—and sent a gallon of gasoline off to a commercial testing laboratory to prove their point. The CSPI's suspicion is that the only real trade secret is that all commercial gasolines of the same octane rating are essentially interchangeable. The CSPI has also persuaded a public-interest law group, the Natural Resources Defense Council, to sue the Environmental Protection Agency to push for faster removal of lead from gasoline—i.e., at a rate which the agency's own consultants have suggested would be feasible.

Jim Sullivan has worked mostly to assist the hundreds of highway action groups which have sprung up nationwide in opposition to urban expressway projects. He has put these groups in contact with experts who can testify for them at hearings and has pressed the Department of Transportation to upgrade its standards for environmental impact statements on these projects.[28] Sullivan seems to be the CSPI's chief entrepreneur, and in 1973 he began a weekly radio program, "Watch-Dog," on a local Washington station with the hopes of syndicating it if it succeeds. In January 1974 he established a public interest science newsletter.

The first-year budget of the CSPI was $20,000, and in the second year it rose to $55,000. Some of this money has been foundation grants, and other money has come in the form of contracts for specific projects (e.g., $10,000 from the Consumers Union for the gasoline-additive project). As the budget has grown, so has CSPI. As of 1973, the full-time staff numbered six, and the center had a regular program for summer science interns.

We have touched on only a few of the CSPI activities. Their scientists are in continual demand for testimony at Congressional hearings, and they have set up a clearing house, Professionals in the Public Service, which puts citizens' groups in touch with appropriate Washington, D.C.-area professionals available for public interest work. Altogether the Center for Science in the Public Interest represents a truly inspirational example of the possibilities of public interest science as a profession.

Conclusion

We have seen in this chapter—and in the entire book—how individual public interest science efforts have appeared in almost every possible institutional framework, and already produced exciting results. But a few robins do not make a spring: the scale of the current public interest science effort is not yet anywhere near commensurate with the challenge posed by technology to our society. Is this movement an echo out of America's individualistic past? Or can it be the seeds of a fundamental transformation of the relationship between scientists and society? It is to these questions which we turn in the next two chapters.

NOTES

1. The NSF later explained that *Environment* had not received the requested support because NSF "was not intended to support activities directed toward applications of science in specific social problem areas but rather to science in general." Quoted in U.S., Congress, Senate, Committee on Government Operations Hearings, *Advisory Committees,* October 6, 1971, Part 3, p. 785.

2. Ralph Nader, "Introduction," *Whistle Blowing,* Ralph Nader, Peter Petkas, and Kate Blackwell, eds. (New York: Grossman, 1972), p. 4.

3. Quoted in *Ibid.,* p. 186.

4. See "A. Earnest Fitzgerald" in *Ibid.,* pp. 39-54. On Fitzgerald's reinstatement, see the *New York Times,* September 19, 1973, p. 1.

5. Quoted in H. Peter Metzger, *The Atomic Establishment* (New York: Simon and Schuster, 1972), p. 258.

6. Bertrand Russell, "The Social Responsibilities of Scientists," *Science* 131 (1960): 391.

7. See, for example, Spencer Klaw's description of this life in *The New Brahmins* (New York: William Morrow, 1968).

8. Quoted in Phillip N. Boffey, "AAAS: is an Order of Magnitude Expansion a Reasonable Goal?," *Science* 172 (1971): 656. (This is the last article in a three-part series on the history of the AAAS. The previous articles begin on pages 453 and 542, respectively.)

9. American Association for the Advancement of Science, Committee on Science in the Promotion of Human Welfare, "Science and Human Welfare," *Science* 132 (1960): 71.

10. U.S., Executive Office of the President, Office of Science and Technology, *Restoring the Quality of the Environment,* Report of the President's Science Advisory Committee (Washington, D.C.: Government Printing Office, November 1965).

11. The information in these paragraphs is taken from Quigley's talk at the Conference on Scientists and the Public Interest: the Role of the Professional Societies, Alta, Utah, September 7-9, 1973. The final report of the ACS task force: American Chemical Society, Committee on Chemistry and Public Affairs, *Cleaning Our Environment: the Chemical Basis for Action* (Washington, D.C.: American Chemical Society, 1969). The direct cost of preparing the report was $67,000, not including ACS staff work and other overhead.

12. Some professional societies have also assumed this helpful but restricted role. For example, the Federation of American Societies for Experimental Biology (an association of six of the more prestigious biomedical professional societies with a collective membership of 13,000) contracted with the FDA in 1973 to convene some twenty-five ad hoc review panels to review the potential hazards of the substances on the FDA's "Generally Recognized as Safe List" of food additives.

13. Both citizens' groups and public officials have complained about their problems in getting scientific advice and assistance. For example, former New York Representative Richard Ottinger spent two years trying to locate a scientist who would testify regarding the effect of the proposed Storm King power plant on a bass spawning ground in the Hudson River. (See Constance Holden, "Public-Interest Advocates Examine Role of Scientists," *Science* 175 (1972): 501.) And the chief deputy attorney general of California complained in 1969 that he was unable to find university petroleum engineers who would testify for the state in its damage suit against four oil companies in connection with the Santa Barbara channel oil leak. (See John Walsh, "Universities: Industry Links Raise Conflict of Interest Issue," *Science* 164 (1972): 411.) A number of state governments have set up science advisory committees, but these have turned out to be mostly pro forma. (See Harvey Sapolsky, "Science Policy in American State Government," *Minerva* 9 (1971), p. 322.)

14. Referendum submitted to members of the Biophysical Society by Peter H. von Hippel, president-elect, winter 1972.

15. "Progress Report on the Development of the Science Advising System . . . ," from Peter H. von Hippel, president, to the members of the Biophysical Society, summer 1973.

16. The conference was sponsored by the American Academy of Arts and Sciences, Western Center, in cooperation with the University of Utah's Engineering Experiment Station, Engineering Department, and Physics Department. The conference was organized by the authors with Barry M. Casper (chairman-elect of the American Physical Society Forum on Physics and Society), under the chairmanship of Peter Gibbs, Chairman of the Physics Department of the University of Utah. Copies of the *Report of the Alta Conference on Scientists in the Public Interest: The Role of the Professional Societies* are available from the American Academy of Arts and Sciences, Western Center, Center for Advanced Study in the Behavioral Sciences, Stanford, California.

17. Ralph Nader, "Introduction," Nader, Petkas, and Blackwell, *Whistle Blowing,* pp. 4-7.

18. *Code of Ethics for Engineers* of the National Society of Professional Engineers (January 1971), quoted in Appendix B of Nader, Petkas, and Blackwell, *Whistle Blowing,* p. 258.

19. *Chemist's Creed,* Approved by the Council of the American Chemical Society, September 14, 1965, quoted in Appendix B of Nader, Petkas, and Blackwell, *Whistle Blowing,* p. 261.

20. For a fictionalized rendition of this situation see Louis V. McIntire and Marion Bayard McIntire, *Scientists and Engineers: the Professionals Who Are Not* (Lafayette, Louisiana: Arcola Communications Co., 1971). The hero of the book, Marmaduke Glum, is a chemist working for Logan Chemical Company. Apparently the book was not sufficiently fictionalized because Louis McIntire, chemist, working for DuPont, was fired after his book came out.

21. Peter Edmonds, Alan Nixon, Peter Petkas, and Frank von Hippel, "Report of the Alta Task Force on Professional Responsibility," *Report of the Alta Conference on Scientists in the Public Interest,* p. 58.

22. Ralph E. Lapp, *Voyage of the Lucky Dragon* (New York: Harper and Bros., 1957).

23. See, e.g., Ralph E. Lapp: *The Weapons Culture* (New York: W. W. Norton, 1968); *Arms Beyond Doubt: the Tyranny of Weapons Technology* (New York: Cowles, 1970); *The Logarithmic Century* (Englewood Cliffs, N.J.: Prentice-Hall, 1973).

24. Alice Kimball Smith, *A Peril and a Hope: the Scientists' Movement in America 1945-47* (Chicago: University of Chicago Press, 1965).

25. James A. Fay and James J. MacKenzie, "Cold Cargo," *Environment,* November 1972, p. 21.

26. Quoted from the first issue of *Center for Science in the Public Interest Newsletter,* April 1971, p. 1.

27. Michael F. Jacobson, *Eater's Digest: the Consumer's Factbook of Food Additives,* (Garden City, N.Y.: Doubleday, 1972).

28. James B. Sullivan and Paul A. Montgomery, "Surveying Highway Impact," *Environment,* November 1972, p. 12.

CHAPTER 18

Congress
and Technology

> *It is the proper duty of a representative body to look diligently into every affair of government and to talk much about what it sees. It is meant to be the eyes and voice, and to embody the wisdom and will of its constituents. . . . The informing function of Congress should be preferred even to its legislative function.* [1]
>
> —Woodrow Wilson

The executive branch by itself cannot be entrusted with ascertaining the general public interest. For one thing, it has its own interests to look after; for another, the access of outside interests to it is too unequal. The framers of the U.S. Constitution were well aware of the potential abuse of executive power—indeed, the Declaration of Independence, written eleven years earlier, had focused on the oppressive acts of King George III. The Constitution, therefore, specifies that establishment of the federal government's basic priorities is the responsibility of a more open and accessible branch of government, a representative Congress. Hence the standard answer to citizen complaints: "Write your Congressman."

But the citizen who does write his Congressman knows that, except for easily remedied personal problems such as an overdue Social Security check or an administrative mistake regarding veterans' or Medicare benefits, he can usually expect little more than soothing reassurances to the effect that the Congressman shares his concern and is keeping a watchful eye on the situation.

Congressional Committees

The problem is that individual Congressmen have very unequal shares of responsibility for overseeing government activities, and those to whom the responsibility has been delegated are usually strongly committed to the status quo. Except on issues currently in the spotlight of national attention, Congress almost always goes along with the recommendations of its committees and subcommittees, whose organization largely parallels that of the executive agencies. And like the federal agencies, the Congressional committees have to a large extent become captives of special-interest groups. Thus Harold Seidman in his book *Politics, Position, and Power* notes that in the Ninetieth Congress (1967-1968) at least half the members of the House and Senate Agriculture committees

were actively engaged in agriculture or related occupations . . . [and] 28 of the 33 members of the House Merchant Marine and Fisheries Committee came from port districts which have a major interest in ship construction and maritime subsidies. Membership on the House and Senate Interior Committees was predominantly from the Western states where reclamation projects, grazing, timber, and mineral rights are issues of primary voter interest.[2]

Seidman then concluded, almost unnecessarily, that parochialism in the executive agencies reflects and is supported by parochialism in their oversight committees."[3]

This parochialism, the existence of which is of course quite natural and unsurprising, goes a long way toward explaining why Congressional committees so often do not take the initiative and may even resist the development of independent information and analyses in their areas of responsibility. Instead, they seem ordinarily to be content to obtain their information from executive agency spokesmen and from the lobbyists for special interests. This is particularly true in complicated technical areas. In evaluating weapons systems, for example, Congress has traditionally obtained most of its information from the military—dismissing most other sources as unqualified. Similarly, in assessing controversies over the side effects of agricultural chemicals, Congress until recently relied almost exclusively upon the chemical industry and the Agriculture Department. In view of this situation, it is perhaps a fortunate by-product of the complexity of modern society and the power of modern technology that an increasing number of problems have ramifications which overlap the jurisdiction of several Congressional committees. (Witness the numerous Congressional hearings in recent years on different aspects of the "energy crisis.") This increases the probability that there will be at least one Congressional committee which will be both competent and sufficiently free of vested interests to provide a fair hearing on any particular technological issue—as did the Senate Foreign Relations committee in the ABM debate after the Senate Armed Services Committee had failed to listen to the ABM's opponents.

CONGRESSIONAL HEARINGS

Congressional hearings can be superb vehicles for bringing a problem to life and dramatizing it. Representative Fountain's grilling of the FDA Administrators on their handling of the cyclamates issue (Chapter 7), for example, had some elements of high drama. The record reveals how the integrity of the FDA bureaucracy was eroded by years of accommodation to the politically potent drug industry. Similarly, the dramatic confrontations between Senator Fulbright and a series of high Defense Department officials (Chapter 5) showed how little importance was actually assigned to technical considerations in the department's "technical" reviews of the ABM system. Thus Congressional hearings provide a unique opportunity to find out how government bureaucracies really operate behind their carefully cultivated public images.

Congressional hearings can also give Congress and the public access to the "experts." There are few scientists who would refuse the invitation of a Congressional committee to testify. Consequently, if the committee is able to determine who the experts are, it can lay before Congress and the public information and analyses which would otherwise just not be available. Panofsky's testimony on the Safeguard ABM system and Garwin's testimony on the SST made unique contributions toward the crystallization and focusing of the issues involved in these debates.

If these are the strengths of the hearing process, it has its weaknesses, too. The quality of a hearing is extremely dependent on the preparation, abilities, and intentions of the Congressmen and staff who choose the witnesses and formulate the questions which are addressed to them. The Congressmen and the staffs do not ordinarily have a technical background: only two Congressmen in the Ninety-third Congress (1973-1974) had an advanced scientific or engineering degree,[4] and there are only a few doctorate-holding scientists on the permanent staffs of individual Congressmen or of Congressional committees. Consequently, the preparation for a hearing tends to be a rather hit-or-miss affair.

Even when the "experts" on each side have presented their arguments, the technical complexities of the issues may so overwhelm the committee that the hearing ends up having only the appearance of a confrontation. Many Congressmen would like to reduce the issue in such debates to one of "my expert is bigger than your expert." But in fact, experts on different sides of an issue usually do not directly contradict each other's statements. Instead, each focuses on that information and those considerations which support his case. And since the witnesses address themselves to the Congressional committee rather than to other experts, it is quite easy for them to talk past one another. In the antiballistic missile debates, for example, the scientist proponents tended to emphasize the hostile intent of the Soviets and Chinese and the consequent requirement for some sort of missile defense, while the opponents argued that the proposed ABM system would be virtually useless against a serious attack. Did this mean that the ABM proponents were unable to rebut the technical criticisms of the opponents? Or that the opponents conceded the need for much greater efforts toward reducing the damage which an enemy could inflict on the United

States with nuclear weapons? Partial answers to these questions were eventually offered during the two-year-long ABM debate.

Most Congressional debates are not so lengthy, however, and such questions would ordinarily be left to the Congressmen and their staffs to struggle with alone. It seems quite likely that, lacking the additional information and analyses which they need to answer these questions, most Congressmen would leave them unresolved and make their decisions on other grounds.

IMPROVING CONGRESSIONAL HEARINGS

Actually, it is not logic but tradition which dictates that witnesses at a Congressional hearing not question each other—as opposed to what happens, in effect, during the adversary proceedings in a courtroom. Perhaps Congressmen enjoy their roles as interrogators. If they could be persuaded to relinquish this prerogative occasionally, however, the payoff might be substantial. Consider the following brief exchange between two experts which occurred in 1957 at a hearing of the Joint Committee on Atomic Energy. Ralph Lapp, having been permitted to present a question from the audience, took issue with a statement by Merril Eisenbud, an AEC official, to the effect that fallout from thermonuclear bomb tests could be increased a millionfold and still be safe. Lapp asked for the radiation dosage in the Troy-Albany area in New York State after the April 1953 nuclear blast in Nevada.

MR. EISENBUD: I would personally estimate it at about ten milliroentgen.
DR. LAPP: Is it proper for me to respond? I have done a little arithmetic. Let
  us take ten milliroentgens, as Mr. Eisenbud estimates, and we multiply [by a
  million] . . . that would be . . . ten thousand roentgens.
SENATOR [CLINTON] ANDERSON [D.-N.M.]: Ten thousand roentgens
  would kill everybody in sight!
MR. EISENBUD: Yes.
SENATOR ANDERSON: So that would mean there would not be any
  immediate danger if you kill everyone in sight?[5]

Even if Congress managed to organize more real debates on technical issues and fewer soliloquies, there are certain deficiencies inherent in the hearing process itself which limit its usefulness as a means of gathering information and advice on technical subjects. Besides the difficulties already mentioned of preparing for the hearing and finding witnesses who are at the same time well informed and reasonably unbiased, there is the more fundamental problem that it is often impossible for any expert, or even a group of experts, to discuss complex issues adequately even among themselves without considerable previous opportunity for study of the relevant information—an opportunity not usually available when an invitation to testify is received. A mechanism is required which will allow extensive investigation and analysis of activities and policies in technical areas so that the issues can be clarified before hearings are scheduled.

Before making up his mind whether to go forward with the development of the Boeing supersonic transport plane, President Nixon commissioned detailed studies from several panels of experts. Such a procedure is routine for major decisions in the executive branch. Yet almost never have Congress and the public been given an authoritative assessment of the costs and benefits of a proposed new technology. Instead, the executive agencies present Congress with a sales pitch, and only rarely does a Richard Garwin or a Matthew Meselson step forward to organize the arguments on the other side.

A recent collaboration between the California legislature and the Rand Corporation (a well-known private "think-tank"[6]) provides a model for a more rational organization of legislative effort in Congress. Like many other states, California has been troubled in recent years by controversies over the siting of new nuclear-energy electric power plants. New state legislation seemed desirable. Before proceeding to draft such legislation, however, the Planning and Land Use Committee of the California State Assembly arranged with Rand for a detailed study of the issues involved. The resulting report, *California's Electricity Quandary*, occupies three summary volumes with more than a dozen supplementary reports.[7] It agrees with the Union of Concerned Scientists that nuclear reactors might not be as safe as the AEC has claimed and suggests that suitable sites might not be available for the sixty additional new nuclear power plants projected by the California utility companies before the year 2000. Finally, as a partial solution to the resulting quandary, the report suggests that significant steps to slow the growth rate of electric power demand in California are feasible.

The report was presented in a private briefing to the chairman of the State Assembly committee, and then it was released to the public in a full-scale press conference at Rand headquarters in Santa Monica. (It was Rand's first press conference.) Next, the report was presented to the entire Planning and Land Use Committee in a major public hearing. A subcommittee then organized several weeks of hearings based on the Rand report, including testimony from California power companies and state agencies. The hearing on nuclear reactor safety featured Henry Kendall, Dan Ford, and officials of the AEC. Finally, the subcommittee chairman, Charles Warren, prepared a bill based on all of this information and discussion. This bill, the Omnibus Energy Conservation and Development Act of 1973, passed the Assembly without modification and was sent to the State Senate. There confusion reigned: twenty-five different energy bills were being considered in the usual piecemeal fashion. The bill that eventually passed the state Senate was a power-plant siting proposal introduced by Senator Alfred Alquist, the chairman of the Senate Committee on Public Utilities and Corporations—but this bill stood no chance of passing the Assembly. The impasse was broken when Senator Alquist agreed to drastically amend his bill to resemble the Warren bill; and the revised Warren-Alquist bill handily passed both houses of the California legislature on September 14, 1973, over strong opposition from the electric utilities—only to be vetoed by Governor Reagan. As of this writing, however, legislative pressure remains strong for repassage and enactment of the Warren-Alquist bill without substantial amendment. Meanwhile, similar bills have been introduced into the legislatures of some half-dozen states.[8]

It would probably be neither practical nor desirable for the U.S. Congress to follow such an elaborate procedure for each of the hundreds of bills it passes each year. But in legislating on complex technical issues, Congress could certainly afford occasionally to adopt a little more rationality in this direction.

The Office of Technology Assessment

Lets face it, Mr. Chairman, we in the Congress are constantly outmanned and outgunned by the expertise of the Executive agencies. We desperately need a stronger source of professional advice and information more immediately and entirely responsible to us and responsive to the demands of our own committees.[9]

— Representative Charles Mosher (R.-Ohio)

Congress has answered the need expressed in the above passage by creating for itself a new institution, the Office of Technology Assessment (OTA), which began operations in late 1973.[10] While it is easy to overrate the impact that the OTA will have on an institution whose nature is still basically feudal, the mere existence of the OTA creates possibilities which would have been dismissed as visionary in the recent past.

What the OTA does is provide for Congress what the President had until recently in the Office of Science and Technology and its Presidential Science Advisory Committee. In the words of the 1972 Technology Assessment Act (Pub. L. 92-484) creating the OTA: "the basic function of the Office shall be to provide early indications of the probable beneficial and adverse impacts of the applications of technology."[11] This is what is called a "technology assessment." A technology assessment can range anywhere from a brief report on a specific technological question to a large-scale study like Rand's report on *California's Electricity Quandary*. (Most routine queries that require only library research will continue to be handled by the Congressional Research Service.[12])

In 1973 ex-Representative Emilio Dadderio was appointed the first director of the OTA. Dadderio had, as a Congressman, nursed the OTA proposal to maturity, before he resigned to run unsuccessfully for governor of Connecticut. The office will eventually have a full-time staff of about twenty professionals. For studies requiring outside resources, however, the OTA is limited only by its appropriations. It is empowered to

enter into contracts or other arrangements as may be necessary . . . with any agency . . . of the United States, with any State, . . . with any person, firm, association, corporation, or educational institution . . . [and] to accept and utilize the services of voluntary and uncompensated personnel . . . and provide [for their] transportation and subsistence.[13]

Besides contracting with major universities and private "think-tanks" for technology assessments, the OTA can develop a mechanism for citizen feedback by requesting studies from organizations like the Center for Science in the Public Interest. The explicit provision for the expenses of volunteers should also encourage all sorts of informal relationships by which individual scientists could contribute important information and analyses. For example, the OTA might appoint several well-qualified monitors, representing a range of viewpoints, who would closely follow the course of an assessment after it had been contracted out and make suggestions to the assessment team and the OTA staff.[14]

STRUCTURE OF THE OTA

Unlike the other two Congressional information services, the General Accounting Office and the Congressional Research Service, the OTA is supervised by what amounts to its own joint Congressional committee, the Technology Assessment Board (TAB). The board consists of six Senators and six Representatives, equally divided between the Democratic and Republican parties.[15] Senator Edward Kennedy (D.-Mass.) was elected the board's first chairman, to serve until January 1975. The Technology Assessment Board can give the conclusions of the OTA reports public visibility and political impact—for example, by holding hearings. It will hopefully also help to protect the OTA from attacks on its appropriations by irate Congressional potentates to whom some of its findings may be unwelcome. And, in cases where the OTA is not receiving cooperation, the TAB is empowered to issue subpoenas.

The OTA will have little impact in the long run, however, unless its work is taken seriously by the technical community. The quality control of the OTA's reports will be partly the responsibility of a part-time Technology Assessment Advisory Council made up of the Comptroller General (who heads the General Accounting Office), the director of the Congressional Research Service, and ten "public" members "to be appointed by the Board, who shall be persons eminent in . . . the physical, biological, or social sciences or engineering or experienced in the administration of technological activities."[16] Additional ad hoc panels may also be appointed to review specific technology assessments or to prepare reports on technical issues relevant to particular pieces of legislation.

By accident, the Office of Technology Assessment was born just as the last remnants of the Office of Science and Technology were being casually swept out the back door of the Executive Office Building. Which raises the question: Will the fate of the OTA be any happier than that of the late OST? In many respects the prospects of the OTA are brighter. In the first place, in contrast to the posture of the Office of Science and Technology, which had only one client—the President—the demands for the services of the OTA will originate from many sources. The chairman, the ranking minority member, or the majority of the membership of any Congressional committee may ask for a study, as may of course the Technology Assessment Board itself or the Director of the OTA "in consultation with the Board." Furthermore, the OTA will constitute the main technical resource of Congress, while the President has always had available the full resources of the entire executive branch—if he trusts them. The priorities in

Congress are sufficiently pluralistic that it can be expected there will always be some Congressional committees to which the OTA will be important at any particular time.

A final advantage of OTA over the late Office of Science and Technology is that it is located in a much more open and public branch of the government. Hopefully the procedures of the OTA, the openness of the work of its experts, the protections suggested above against bias in their reports, and the open publication of these reports for public use and criticism will set an example which the executive branch will be obliged to follow. Given a choice, it is probable that many scientists would prefer to work under such conditions. And their reports are much more likely to obtain full consideration in Congress—and the executive branch as well—if they are openly available. Recall that the public release of the Rand report *California's Electricity Quandary* generated a great deal of press attention, which in turn helped to lubricate the California legislative machinery. Indeed, the OTA should establish mechanisms for the information and involvement of the larger public in its activities—at least a newsletter to publish announcements of proposed new technology assessments, progress reports, and brief accounts of completed assessments. Of course, some confidentiality will be necessary on occasion to protect military security and industrial trade secrets. In these cases the damage done to open public debate can be minimized by publishing "sanitized" reports containing the OTA's unclassified analyses, conclusions, and recommendations, omitting only the technical details being protected.

Getting Congressional Attention

The location of the OTA in Congress gives it many advantages, but there are also obvious disadvantages. Former Senator Joseph Clark did not express an uncommon view when he described Congress as "the sapless branch."[17] Congress has traditionally deferred to the executive branch on technological matters. The resources made available by the OTA will enable Congress to challenge the executive branch more easily in these areas—but there is little basis in recent history to believe that Congress will rise to the occasion without a great deal of prodding. Any resemblance between most Congressional committees and a group of Nader's Raiders is purely coincidental.

Despite its front-row seat on the operations of the federal government, Congress raises few issues of a nonparochial nature on its own initiative. It seems that Congressmen are usually just too busy servicing the needs of their own political constituencies to have much time or energy left over to worry about the general public interest. It requires political skill to get a Congressman's attention and support.

In all the cases that we have discussed the basic ingredient which attracted

Congressional attention was an aroused public. This is particularly true in the big debates: those over the SST and ABM. It was public concern over the sonic boom that originally triggered the major Congressional debates over the SST, and it was the suburban opposition to "bombs in the backyard" that revitalized the ABM debate. It is true that, after these beginnings, the Congressional debate branched out into other problems relating to these two technologies—but it was the public outcry that originally drew Congressional attention. Such national debates provide Congressmen with an audience. And with national news coverage focused on them, Congressmen are more likely to take the issues seriously.

Besides their natural sensitivity to publicity, there is a deeper reason why Congressmen respond much more attentively to an issue which has already received a great deal of public debate than they do to an issue of similar merit which comes to Congress unheralded. On controversial issues, Congress does not actually *decide*; rather, it *ratifies* what it takes to be the popular will. Thus, for example, in the development of the labor movement in the United States, years of labor organizing, strikes, and sometimes violent controversy preceded the eventual passage of the Wagner Act in 1938.[18] Similarly, Congressional action finally cutting off funds for the bombing of Cambodia in 1973 came as a much-delayed anticlimax to general public disaffection with the war in Indochina.

Of course, few technological issues generate political struggles as fierce as those which surrounded the ABM and the SST. Fortunately, most issues—like the cyclamates issue or the dangers of cross-country transportation of nerve gas—can be handled at a lower level of confrontation. In cases such as the latter, however, it is still useful to represent a political constituency which the Congressman being approached takes seriously, or to be introduced by an individual whom he respects, or to have already attracted news media attention to the issue. It is also almost essential to develop the issues for him and his staff with clear and persuasive written arguments so that they may choose which ones they wish to use for their own purposes.

Keeping Congressional Attention

Perhaps the most important problem that the concerned citizens' group faces, once it has first engaged Congress's attention, is keeping it. Elizabeth Drew has described the problem as follows:

The people in Congress, like people who are not in Congress, are endowed with a rather limited attention span. A member of Congress' relationship with any particular national issue is likely to be of rather brief duration. Anyone who stays with an issue for very long may be considered by his colleagues and by the press to be a little bit odd, somewhat obsessive, a joke. (They laughed at the way Wayne Morse went on about the [Vietnam] war.)[19]

This is why the most effective weapon in the arsenal of the defenders of the status quo is delay.

In order to focus continued Congressional attention on questions relating to the general public interest rather than to special interests, it helps if there is action in other arenas. As we have already remarked, Congressmen like an audience for their efforts—but most Congressional hearings are ignored unless they are coupled with public or legal controversies that have already drawn media attention. The battle over DDT provides a prime example of how an issue was kept alive over the years by the action shifting continuously from one arena to another: first Rachel Carson's *Silent Spring*; then the report of the President's Science Advisory Committee; then the local courts and the state legislatures; new findings by scientists on the pervasiveness and toxicity of DDT; administrative hearings in front of the Environmental Protection Agency; more advisory reports; more court actions; etc. The Office of Technology Assessment should deliberately try to compensate for the spasmodic nature of Congressional, public, and even executive-branch attention by undertaking periodic reviews of a variety of issues such as pesticide usage, nuclear reactor safety, or land-use planning—whether these areas are currently the focus of controversy or not—with reports to Congress on its findings. In this way Congress and the public could find out what impact previous legislation has actually had and be warned of new problems before they reach crisis proportions.

Another way in which to keep Congressional attention is of course to emulate the special interests and become involved in Congressional elections. Various groups involved in debates over technology have done just this. Meselson approached several Congressmen with the chemical and biological warfare issue through their big campaign contributors. In the SST debate many local anti-SST groups inserted the issue into Congressional campaigns. And at least two public interest groups have dedicated themselves with considerable success to using the electoral process to change Congress so that it will become more favorable to their views. One, the Council for a Livable World, contributes to the political campaigns of Senatorial candidates from small states who favor its arms control objectives; the other, the Friends of the Earth's League of Conservation Voters, before each national election issues a list of a "dirty dozen" Congressmen whom it would most like to see defeated.

Congressional Staff

Lack of time, lack of staff, lack of expertise, pitted against the Pentagon's legions of experts, frustrated our [i.e., the Senate Armed Services Subcommittee on Research and Development's] attempts to make a significant number of line item cuts. Ultimately we had to resort in the main to asking the Executive Department to make percentage cuts, instead.

Most every item should be carefully considered and closely challenged. But until Congressional committees charged with this responsibility have adequate staffs, skilled in investigation and interrogation, we will not be able to meet this charge. We will have no alternative but to continue with percentage cuts, thereby relinquishing to the Executive branch the real decision-making power.[20]

— Senator Thomas J. McIntyre (D.-N.H.)

It cannot be overemphasized that it is a Congressman's staff which represents his memory and his ability to follow through on an issue. The staff member has more time than the Congressman to listen to arguments, and once he understands and is convinced by them, he is likely to know which ones will be persuasive to his boss. Persuading a key staff member of the importance of an issue and educating him on what must be done may therefore be as important as persuading the Congressman himself—or even tantamount to it. Furthermore, the Congressman is more likely to be willing to commit his staff man to the fray if that staffer is already well informed and chomping at the bit.

Each Representative has a staff of about eight people in his Capitol Hill office, and each Senator's Congressional staff numbers about twenty; in addition, each of the thirty major Congressional committees has a staff of about twenty-five. These numbers may at first sight seem rather large, but most of the Congessmen's personal staff is concerned with political or office chores—case work, answering constituent mail, and the like. A Congressman's Legislative Assistant and Administrative Assistant are in charge of Congressional business and running the office, respectively. Each member of Congress thus has at most a few staff members who can afford to specialize in areas of special interest to him—unless he happens to chair a subcommittee or, better yet, a major committee. But even committee staffs comprise mostly lawyers and political types. Consider the Senate Commerce Committee, for example. Its eight subcommittees are responsible for aviation, communications, consumer affairs, environment, foreign commerce and tourism, merchant marine, oceans and atmospheres, and surface transportation; and they oversee the functioning of the Department of Commerce (including the National Bureau of Standards, the National Oceanic and Atmospheric Administration, and the Patent Office), most of the Department of Transportation, and four federal regulatory agencies: the Federal Aviation Administration, the Federal Communications Commission, the Federal Power Commission, and the Interstate Commerce Commission. Yet with all this technology under its supervision, the Senate Commerce Committee has only *one* staff specialist with an advanced degree in engineering or science. Other committees with jurisdiction over science and technology are in a similar position, as Senator McIntyre's lament, quoted at the beginning of this section, attests.

It is obvious that Congress is woefully understaffed with technical expertise. Recognizing this, a number of professional societies have recently initiated a Congressional Scientist-Fellow Program, whose purpose is to place outstanding younger scientists and engineers on Congressional staffs for approximately one year. The first scientist-fellow, Barry Hyman, a mechanical engineer, began working with the Senate Commerce Committee in January 1973.[21] During his one-year fellowship he helped draft and organize hearings on three major bills. In September 1973 he was joined by six additional scientist-fellows: two electrical engineers, two physicists, a molecular biologist, and an assistant dean on leave from Yale Medical School. Congress appears to desire the services of many more such fellows: the American Association for the Advancement of Science, which is coordinating the program, has received some eighty requests for scientist-fellows from Congressmen, and the competition among Congressmen was very hot to see who could sign up the first fellows. Additional professional societies were expected to join in sponsoring the Congressional Scientist-Fellow Program in 1974, and foundation support was being sought which would allow a considerable further expansion.

Hopefully the presence of these scientists on Congressional staffs will increase the willingness of Congressmen to venture into the technology policy area. Congressmen may even begin to seek scientific staff with their own funds. Indeed, all of the first group of congressional Scientist-Fellows have been invited to stay on as staff members—and about half have decided to accept. There is a precedent for this: the two permanent Congressional staff members with doctoral degrees in physics originally came to Congress with outside support. One, Tom Ratchford, a physicist on the staff of the House Science and Astronautics Committee, first came to work for this committee under the Congressional Fellowship Program of the American Political Science Association. The other, John Andelin, a physicist who is now Administrative Assistant to Representative Mike McCormack (D.-Wash.), initially came as a volunteer.

Those Congressional Scientist-Fellows who return to universities and industry also can have a great impact on the relationship between Congress and the scientific community. They can be points of contact for Congressional staff searching for experts and information on particular issues. With their knowledge of how to get important issues and information to the Congressmen and Congressional committees where it will do the most good, they can be extremely useful to those scientists involved in public interest science activities in their home institutions.

Conclusion

In summary, citizens should think of Congressmen not as champions to be enlisted in the cause, but as a distracted, reluctant, and skeptical audience that sometimes can be persuaded to pass remedial legislation or to put pressure on a wayward government agency—once some group of citizens has developed the case and put it before the public or the courts. This prospect may appear rather forbidding, but sometimes Congress is the only resort. Even efforts which are only partially successful can make Congress and the public more sensitive to an issue when it arises again. In the meantime the new Office of Technology Assessment and the Congressional Scientist-Fellow Program should significantly increase Congress's ability to recognize and deal with technological issues.

NOTES

1. Woodrow Wilson, *Congressional Government* (1885); quoted in U. P. Harris, *Congress and the Legislative Process* (New York: McGraw-Hill, 1967), p.4.

2. Harold Seidman, *Politics, Position, and Power: The Dynamics of Federal Organization* (New York: Oxford University Press, 1970), p. 40.

3. Ibid.

4. Representative Mike McCormack (D.-Wash.) has a master's degree in chemistry from Washington State University and Representative James G. Martin (D.-N.C.) has a Ph.D. in chemistry from Princeton. Two senators, James G. Abourezk (D.-S.D.) and Dewey F. Bartlett (D.-Okla.), and several Representatives have bachelor's degrees in engineering.

5. *The Nature of Radioactive Fallout and Its Effects on Man*, U.S. Congress, Joint Committee on Atomic Energy, 85th Cong., 1st sess., May-June 1957; quoted in H. Peter Metzger, *The Atomic Establishment* (New York: Simon and Schuster, 1972), p. 98.

6. "Think-tanks" are organizations which do studies of technical issues, usually for fee-paying clients. Rand, whose name stands for research *and* development, is one of the oldest such organizations. It was founded shortly after the Second World War, mainly to do research for the Air Force. For a popular account, see Paul Dickson, *Think Tanks* (New York: Ballantine, 1972).

7. Ronald Doctor et al., *California's Electricity Quandary* (Santa Monica, Calif.: Rand Corporation, 1972), 3 vols.

8. The information in this paragraph is from Maureen Fitzgerald, "Who Does What in the Energy Crisis," *California Journal*, December 1973, pp. 407-9; Emilio Veranini, chief consultant to the California Assembly Committee on Planning and Land Use, "Political Interaction and Energy Policy" (unpublished talk at American Physical Society meeting, Berkeley, California, December 27, 1973); and from correspondence with Warren and Alquist.

9. Representative Mosher is the ranking Republican on the House Science and Astronautics Committee. Its Subcommittee on Science, Research, and Development, formerly chaired by Representative Emilio Q. Daddario (D.-Conn.), authored the bill creating the Office of Technology Assessment. The quoted remark was made in support of this bill during floor debate (*Congressional Record* 118 (1972): 3202).

10. For the background of the OTA and various recommendations for its mode of operation, see Anne Hessing Cahn and Joel Primack, "Technological Foresight for Congress," *Technology Review*, March-April 1973, pp. 39-48. The concept of technology assessment was developed in *Technology: Processes of Assessment and Choice*, prepared by the National Academy of Sciences for the House Committee on Science and Astronautics (Washington, D.C.: Government Printing Office, July 1969). See also *Technical Information for Congress*, prepared by the Congressional Research Service for the House Committee on Science and Astronautics, rev. ed. (Washington, D.C.: Government Printing Office, 1971). An amusing introduction to the business of technology assessment is given by Nina Laserson Dunn, "Technology Assessment at the Threshold," *Innovation* 27(1973): 14-27.

11. Technology Assessment Act (Pub. L. 92-484), Sec. 3(c).

12. For a review of the effectiveness of the Congressional Research Service, see Richard E. Cohen, "Information Gap," *National Journal,* March 17, 1973, p. 379.

13. Technology Assessment Act (Pub. L. 92-484), Sec. 6(a)(2) and 6(a)(4).

14. This suggestion is Jessica Tuchman's.

15. In the Ninety-third Congress, the membership of the Technology Assessment Board consisted of Senators Kennedy (D.-Mass.), Hollings (D.-S.C.), and Humphrey (D.-Minn.), Case (R.-N.J.), Dominick (R.-Colo.), and Schweiker (R.-Penn.), and Representatives Davis (D.-Ga.), Teague (D.-Tex.), Udall (D.-Ariz.), Mosher (R.-Ohio), Harvey (R.-Mich.), and Gubser (R.-Calif.).

16. Technology Assessment Act (Pub. L. 92-484), Sec. 7(a)(1). The first ten public members were appointed to the Technology Assessment Advisory Council in late 1973: Harold Brown, J. Frederick Buey, Hazel Henderson, J. M. Leathers, James McAllister, Eugene Odum, Frederick Robbins, Edward Wenk Jr., Gilbert White, and Jerome Wiesner.

17. Joseph S. Clark, *Congress: The Sapless Branch* (New York: Harper & Row, 1964).

18. This point was drawn to our attention by Richard Levin at the 1973 Alta Conference on Scientists in the Public Interest. (See note 8, Chapter 16.)

19. Elizabeth Drew, "Members of Congress are People," *New York Times,* January 29, 1973, p. 29.

20. Press release issued by Senator McIntyre's office, November 7, 1969.

21. The American Society of Mechanical Engineers is sponsoring Barry Hyman in cooperation with his home institution, George Washington University. The American Association for the Advancement of Science, partly out of its own funds and also utilizing a personal contribution from its treasurer, William Golden, is supporting three Congressional fellows; the American Physical Society is supporting two fellows; and the Institute of Electrical and Electronics Engineers is sponsoring one fellow who is supported by his home institution, Georgia Tech. The AAAS is coordinating the program. One of the authors of this book (Joel Primack) was involved in the creation of the Congressional Scientist-Fellow Programs of the AAAS and APS, and the other (Frank von Hippel) served on the 1973 selection committees for both organizations.

PART VI

Conclusion

The knowledge that the public possesses on any important issue is derived from vast and powerful organizations: the press, radio, and, above all, television. The knowledge that governments possess is more limited. They are too busy to search out the facts for themselves, and consequently they know only what their underlings think good for them unless there is such a powerful movement in a different sense that politicians cannot ignore it. Facts which ought to guide the decision of statesmen—for instance, as to the possible lethal qualities of fallout— do not acquire their due importance if they remain buried in scientific journals. They acquire their due importance only when they become known to so many voters that they affect the course of the elections. . . .

. . . What ought to be known widely throughout the general public will not be known unless great efforts are made by disinterested persons to see that the information reaches the minds and hearts of vast numbers of people. I do not think this work can be successfully accomplished except by the help of men of science. . . . I think men of science should realize that unless something rather drastic is done under the leadership or through the inspiration of some part of the scientific world, the human race, like the Gadarene swine, will rush down a steep place to destruction in blind ingnorance of the fate that scientific skill has prepared for it.

—Bertrand Russell, in "The Social Reponsibilities of Scientists," *Science*, February 12, 1960

CHAPTER 19

The Choice for Scientists and for Society

Our political system is currently in a state of flux. Faith in institutions and faith in progress are on the decline. Yet for many the disillusionment is accompanied by a deepened understanding of the importance of the fundamental democratic processes and has led to a new political activism. Only time will tell whether the signs of decay or those of renewal more accurately portend the future.

The manner in which technology is exploited—for whose benefit? at whose expense?—will substantially influence this future. And the case studies in this book show that the ways in which scientists inject information into the decision-making process will to a large extent determine whether future policy making for technology will be made in a secret totalitarian manner or in an open democratic one. Only individual scientists can equip concerned citizens with the information and confidence which they need to answer the government's constant challenge: We have our experts; where are yours?

Will scientists accept their public responsibilities? Or will they largely restrict themselves to the tasks assigned them by their employers—thus accepting the status of supertechnicians and paving the way for ever-greater concentrations of power? The answer to these questions must depend upon the independence of scientists, the encouragement society gives to their public interest activities, and the creativity they and their allies exhibit in institutionalizing public interest science.

As this book is written, the influence of scientists in government and their economic independence are probably lower than they have been since before World War II. The President's science advisory apparatus has been dismantled after being essentially ignored for some years, and the priority of support for science and technology has been downgraded except in a few politically profitable areas. As a result of this decreased support and because of the tremendous increase in their numbers, scientists who ten years ago would have been able to choose from among a variety of attractive research jobs are now often unable to continue in research at all.

There is a natural tendency under these circumstances for scientists to concentrate on the bread-and-butter issues of professional survival. This is reinforced by the increasing tendency of administrators to treat scientists more as ordinary employees who should "get on the team" than as irreplaceable assets who must be humored and coddled lest they be wooed away by better job offers. Thus most scientists are becoming painfully aware that they are no longer a privileged elite and must in the future share the uncertainties and vulnerabilities of ordinary men.

At the same time that the economic independence of scientists has been so reduced, a political atmosphere has developed in which the public seems to be almost begging them for independent information on the possibilities and dangers of technology. After the disastrous involvement of the United States in the Indochina war and in the wake of revelations that government decisions have been "for sale" on an apparently unprecedented scale in exchange for political contributions to the President, the public has become less and less comfortable with the invitation from federal agencies to "leave the driving to us." The Indochina war demonstrated particularly clearly the almost unlimited capacity of a powerful bureaucracy to deceive itself, to avoid making unpleasant decisions, and to mislead the public in the process. In many respects each of the issues discussed in this book—the SST, DDT, nuclear reactor safety, the ABM—has been a technological Vietnam. Sanity had to be forced on the responsible bureaucracy in each case by an aroused public.

The debates over these issues have revealed the great reservoir of citizen interest and the organizing talent and energy available in this country—once the issues have been made clear and intelligible. At the same time, however, the past decade of political debate has caused considerable discouragement among these same individuals. The political battles over the issues of racism, the Indochina war, the arms race, and environmental pollution have shown that these issues are much more complex than was thought initially and that there are no easy political solutions or "technological fixes." Each bit of progress has revealed a new layer of interconnections of the problems with our social structure, until it seems almost as if one can solve no specific problem without restructuring the entire society. But few people can indefinitely sustain an intense involvement with issues remote from their personal lives. Sooner or later most of us must withdraw from campaigns to save the world in order to mend fences at home and on the job. Obviously the challenge is to develop goals which are not only realistic but also personally meaningful to large numbers of people.

Currently it takes an unusually adventurous and astute individual to be an effective public interest scientist. Such exceptional personalities are no more common in science than in other fields, and society has become too complex to depend for salvation on the activities of a few individuals. The challenge to scientists and citizens alike, therefore, is to civilize the environment of public interest science so that more scientists can contribute. In this connection it is instructive to study the "opposition": government and corporate bureaucracies.

Bureaucracies provide their members with a very important commodity:

legitimacy. There is a widespread presumption that an individual as a representative of an organization has a legitimate reason to be concerned with an issue affecting his organization, while if the same individual takes up an issue on his own, the presumption is that he is a crackpot. The flimsiest sort of organizational base can have a substantial effect in raising the debate above the level of personalities. Thus, for example, Henry Kendall and Dan Ford as representatives of the Union of Concerned Scientists—an organization little more substantial than its irregularly scheduled meetings, secretary, and post office box—were able to challenge the Atomic Energy Commission on an organization-to-organization basis.

There is no reason to consider it "illegitimate" or "immoral" to exploit such an institutional "front" as a means of precluding distracting debates over the qualifications of the participants and of forcing discussion of the issues themselves. Indeed, if our case studies are any guide, it seems that, despite the great resources of expertise available to government agencies (such as the AEC and FDA), the credentials of agency decision makers and their reasons for making decisions will often stand up under inspection much more poorly than the arguments of carefully prepared public interest scientists. Or to put it another way: If an agency spokesman can invoke legitimacy by virtue of the expertise at the disposal of his agency, why should not the public interest scientist also be allowed to claim legitimacy by virtue of his affiliation with a university, a scientific society, or a public interest group? Once it has been established that neither a government agency nor its challenger has an exclusive monopoly on truth or good judgment, the debate can focus on the merits of the case made by each.

In fact, as more young scientists become involved in public interest activities, as the issues multiply, and as legal tools are developed making policy-making for technology subject to judicial intervention, public interest science is finding a home in a great variety of organizations. In the past the issues were brought into the public arena when extraordinary individuals with a public identity raised their voices: Rachel Carson (pesticides), Linus Pauling (radioactive fallout), Hans Bethe (ABM). The new public interest scientist has to do much more than raise his voice to get a hearing: Shurcliff (SST) became a fund raising and media expert, Kendall and Ford (nuclear reactor safety) immersed themselves in the AEC's administrative hearing process, Wurster became involved in legal challenges to the use of persistent pesticides at the state and then the national level, Meselson (CBW) became an expert lobbyist with both Congress and the White House. Organizational efforts have grown naturally out of each of these enterprises: the Citizens League Against the Sonic Boom, the Consolidated National Intervenors, the Environmental Defense Fund, and the AAAS Herbicide Assessment Commission.

There seem to be an infinite variety of forms which public interest science can take. The public support exists, scientists want to become involved, and there are plenty of dragons with which to do battle.

APPENDIX

A Summary of Science Advisory Organizations

The President

The position of **President's Science Advisor**[1] established in 1957 by President Eisenhower in response to the challenge of the Soviet Union's triumphantly successful launching of their *Sputnik* space satellites, has been occupied in succession by James R. Killian, Jr. (1957-1959), George B. Kistiakowsky (1959-1961), Jerome B. Wiesner (1961-1963), Donald F. Hornig (1964-1969), Lee A. DuBridge (1969-1970), and Edward E. David (1970-1973). The Science Advisor, a full-time Presidential aide, chaired the **President's Science Advisory Committee** (PSAC), consisting of eighteen scientists and engineers serving staggered four-year terms who met regularly in Washington for two days each month. PSAC members also supervised a number of scientific panels on specialized topics, consisting in all of several hundred scientists. The full-time staff of the Science Advisor, which included a dozen or so scientists, was christened in 1962 the **Office of Science and Technology** (OST). The principal function of the President's Science Advisor, PSAC, and OST was to provide independent advice on technological issues to the President and the Budget Bureau, advice which could serve to check and counterbalance the sometimes self-serving recommendations sent to the White House by the executive-branch agencies.

President Nixon abolished PSAC and OST in early 1973 and transferred some of the responsibilities of the President's Science Advisor to Guyford Stever, director of the **National Science Foundation** (NSF), the principal federal agency charged with the support of pure science. The President also continues to receive science advice from the three-member **Council on Environmental Quality** (CEQ), created by authority of the 1969 National Environmental Policy Act and charged with receiving environmental impact statements and preparing an annual public report, and from the **Atomic Energy Commission** (AEC) and other executive branch agencies.

The Executive Branch

The executive branch departments and agencies most directly concerned with science and technology have large science advisory organizations. In the **Department of Defense** (DOD), the **Director of Defense Research and Engineering** (DDR&E), who ranks just below the Deputy Secretary of Defense, is responsible for administering DOD-sponsored research and development and for coordinating advanced weapons systems. Science advice is given to the Secretary of Defense, through the office of the DDR&E, by the **Defense Science Board** (DSB), whose 24 members are drawn mainly from defense-related industries. Each of the military services also has its own science advisory committee, and there are many additional committees of scientists advising various DOD officials on specialized technical matters. In addition to all of these part-time committees of scientists and engineers, the Defense Department also supports a number of non-profit private "think tanks": the **Institute for Defense Analyses** (IDA) advises the Secretary of Defense, **Rand Corporation** advises the Air Force, etc. IDA's "Jason" division, a group of about 40 prominent academic scientists (mostly theoretical physicists), has been consulting for the Defense Department since 1958. (In 1973, Jason shifted its affiliation to the Stanford Research Institute, another think-tank largely supported by the Defense Department.)

The principal science advisory committees of the **Atomic Energy Commission** are the **General Advisory Committee** (GAC), which was for several years after the Second World War the government's most influential science advisory committee, and the **Advisory Committee on Reactor Safeguards** (ACRS). The AEC, NSF, and **Department of Health, Education, and Welfare** (HEW) all devote several hundred million dollars annually toward sponsorship of research in universities and federal laboratories, including AEC's **National Laboratories** and HEW's **National Institutes of Health**. Each of these agencies has numerous scientific advisory committees and each agency also regularly consults with recognized scientists on the best allocation of funding among competing research proposals (this is called the "peer-review system"). A number of federal departments and agencies that have only in recent years begun to conduct large-scale research and development programs possess somewhat less extensive scientific advisory arrangements, and depend mainly upon the **National Academy of Sciences** for science advice.

Congress

The **Office of Technology Assessment** (OTA), which began operation late in 1973, was established in order to increase Congress' access to competent advice on technological issues. The first director of the OTA is former Representative Emilio Q. Daddario (D.-Conn.), who had seven years earlier first initiated the OTA legislation. A committee of six members each from the Senate and the House of Representatives, known as the **Technology Assessment Board**, acts as a board of directors for the OTA; and there is also an OTA **Technology Assessment Advisory Council**, composed mainly of scientists. The OTA is expected to undertake the study of major unresolved technological issues confronting Congress, with research being performed by universities or private research organizations under supervision of the OTA staff. Congress will also continue to receive assistance in library research on technical issues from the **Congressional Research Service** (CRS) of the Library of Congress.

The National Academies

The 1,000-member **National Academy of Sciences** (NAS) and its smaller offspring the **National Academy of Engineering** (NAE) and the **Institute of Medicine** are largely honorary organizations. However, the Congressional charter of the NAS, adopted in 1863, specifically requires that "the Academy shall, whenever called upon by any department of the Government, investigate, examine, experiment, and report upon any subject of science or art."[2] This advisory obligation is fulfilled mainly through the activities of the NAS's **National Research Council** (NRC), which supervises the work of more than 6,000 scientists and engineers serving part-time on nearly a thousand advisory committees.

The nature of NAS-NRC committees varies considerably. At one end of the spectrum are industry-dominated panels advising the Defense Department on "textile dyeing and finishing," or the Agriculture Department on "dog nutrition." At the other end are groups like the NAS **Committee on Science and Public Policy** (COSPUP), which has prepared thoughtful reports on subjects like the need for technology assessment. In order to prevent further fiascos like the misleading report on sonic boom damage described in Chapter 4, the NAS established in 1971 a special review committee for potentially controversial NAS-NRC reports, chaired by the NAS vice-president. This committee has several times been successful in effecting substantial improvements in Academy reports.

REFERENCES

1. Names of advisory positions and organizations are printed boldface here to make them easier to locate. For more information on advisory organizations, consult the index for relevant page references and footnotes. For a general reference on this material, see Frank von Hippel and Joel Primack, *The Politics of Technology: Activities and Responsibilities of Scientists in the Direction of Technology* (Stanford, Calif.: Stanford Workshops on Political and Social Issues, 1970), which is comprehensive but somewhat out of date; *The Science Committee* (Washington, D.C.: National Academy of Sciences, 1972), 2 vols.; and *Federal Advisory Committees: First Annual Report of the President to the Congress, Including Data on Individual Committees* (Washington, D.C.: Government Printing Office, 1973 and 1974), 4 vols. plus an index.

2. Quoted in *National Academy of Sciences, National Academy of Engineering, Institute of Medicine, National Research Council: Organization and Members* (Washington, D.C.: National Academy of Sciences, annual), p. 10. This publication also includes a list of all NAS-NAE-NRC committees, with their memberships.

INDEX

A

ABM Advisory Committees, 47, 59; Johnson administration, 62-64; Kennedy administration, 61; Nixon administration, 65-71

Advisory Committee Act, 35-36, 113-114, 120

Advisory Committee on Reactor Safeguards, *see* Atomic Energy Commission

American Association for the Advancement of Science (AAAS) (*see also* Herbicide Assessment Commission), 81, 143-145, 154-160, 254-255

American Chemical Society, 146; code of ethics, 259; pollution study, 255; professional relations committee, 260

American Physical Society (APS), 256

Andelin, John, 279

Antiballistic Missile (ABM), 6, 59-62, 178-179, 188-189, 192, 210 accident danger, 186, 192; Boston site; 187-188; Chicago site, 183-187; Congressional hearings, 60-71; Congressional votes, 68, 190, Detroit site, 187; and Joint Chiefs, 60-62, 180; penetrability of defense, 61, 64, 70, 109, 181; Seattle site, 180, 182-183; Soviet system, 62, 181, 189; Strategic Arms Limitations Talks, 69, 190-191

Argonne scientists, 183-187, 191

Arms Control and Disarmament Agency, 145-148

Atomic Energy Commission (AEC) (*see also* Nuclear gas stimulation, Nuclear power plants, Nuclear reactor licensing, Nuclear reactor safety, plutonium)

Advisory Committee on Reactor Safeguards, 113, 120, 210, 214, 218, 229, 231-232, 290; and Freedom of Information Act, 113, 118, 120, 218; General Advisory Committee, 290; National Reactor Testing Station, 211, 213-214, 217, 222-223, 231; nuclear gas stimulation, 171-174; Oak Ridge National Laboratory, 215-217, 220-221, 223; and nuclear test-ban treaty, 40; reorganization (1973), 230-231; Rocky Flats fire, 167-170; underground nuclear testing, 171

Audubon Society, 130, 232, 263

B

Becket, Elise, 202

Bethe, Hans, 61, 64, 109, 180-182

Bionetics Research Laboratories, 74-79, 83-84, 251

Biophysical Society: Public interest roster, 257-258

Bjorken, James D., 201-202

Borlaug, Norman E., 138-139

Bronk, Detlev, 65-66

Brown, Harold, 64

Businessmen for the Public Interest, 211

293